SpringerBriefs in Applied Sciences and Technology

SpringerBriefs present concise summaries of cutting-edge research and practical applications across a wide spectrum of fields. Featuring compact volumes of 50 to 125 pages, the series covers a range of content from professional to academic.

Typical publications can be:

- A timely report of state-of-the art methods
- An introduction to or a manual for the application of mathematical or computer techniques
- A bridge between new research results, as published in journal articles
- A snapshot of a hot or emerging topic
- An in-depth case study
- A presentation of core concepts that students must understand in order to make independent contributions

SpringerBriefs are characterized by fast, global electronic dissemination, standard publishing contracts, standardized manuscript preparation and formatting guidelines, and expedited production schedules.

On the one hand, **SpringerBriefs in Applied Sciences and Technology** are devoted to the publication of fundamentals and applications within the different classical engineering disciplines as well as in interdisciplinary fields that recently emerged between these areas. On the other hand, as the boundary separating fundamental research and applied technology is more and more dissolving, this series is particularly open to trans-disciplinary topics between fundamental science and engineering.

Indexed by EI-Compendex, SCOPUS and Springerlink.

Viktor Koval

Editor

Renewables in the Circular Economy and Business

 Springer

Editor
Viktor Koval
Department of Business and Tourism
Management
Izmail State University of Humanities
Izmail, Ukraine

ISSN 2191-530X　　　　　　　ISSN 2191-5318　(electronic)
SpringerBriefs in Applied Sciences and Technology
ISBN 978-3-031-72173-1　　　　ISBN 978-3-031-72174-8　(eBook)
https://doi.org/10.1007/978-3-031-72174-8

This Springer imprint is published by the registered company Springer Nature Switzerland AG
The registered company address is: Gewerbestrasse 11, 6330 Cham, Switzerland

If disposing of this product, please recycle the paper.

An Overview of Renewables in the Circular Economy and Business

This book presents research into the macroeconomic analysis of renewable energy consumption and the financial implications of energy development. Emphasis is placed on the research into the possibilities of a circular economy and approaches to sustainable energy development using renewable sources.

Chapters explore macroeconomic factors influencing the adoption of renewable energy sources and offer a detailed understanding of how renewable energy sources can be used to optimize resource use, improve energy sustainability and economic security, and improve safety and energy efficiency through ongoing assessment and maintenance, implementation initiatives in the field of sustainable energy based on the introduction of new technologies, and improvement of regulatory relations.

In order to evaluate the contribution of renewable energy to the sustainable growth of the electrical power industry, this evaluation will concentrate on the energy providers and consumers, primarily families and companies that are progressively transitioning to prosumer status in the renewable energy market.

The book will facilitate the creation of a vision aimed at the development of renewable energy power generation systems and the study of systems in these changing climate conditions to find sustainable solutions.

Regional policies can reduce investor risk, which will support commercially successful renewable energy projects. The theory, design, modeling, application, control, environmental concerns, and the future evolution of energy policies are all examined and discussed in this book. Reviews and case studies of the development of renewable energy-based power generation systems and examining those systems in light of changing climate circumstances to find sustainable solutions are also included.

The editor wants to express gratitude to all the contributors and reviewers who have aided in the publication of this book. The editor would also like to thank Springer for their full support during publishing.

Izmail, Ukraine   Viktor Koval

The original version of the book has been revised: The copyright year level have been corrected. A correction to this book can be found at https://doi.org/10.1007/978-3-031-72174-8_7

Contents

Macroeconomic Analysis of Renewable Energy Consumption: A Panel Data Approach

Yevhen Revtiuk⬛, Viktor Koval⬛, and I Wayan Edi Arsawan⬛

Abstract The use of renewable energy sources enables the decarbonization of the economy. Moreover, the share of renewable energy in the global energy balance is growing. Government policymakers encourage increased investment in renewable energy generation capacity, even though the higher price of such energy compared with energy from traditional sources remains an obstacle to the complete abandonment of the use of fossil fuels for energy generation. This study examines the relationship between macroeconomic characteristics and energy consumption from renewable sources in 159 countries from 1990 to 2020. A positive association was identified between economic inequality and renewable energy consumption (a 1% increase in economic inequality was associated with a 0.58% increase in renewable energy consumption) and between GDP per capita and renewable energy consumption (a 1% increase in GDP per capita was associated with a 0.11% increase in renewable energy consumption). The negative association between economic inequality and the GDP per capita interaction with renewable energy consumption mitigates the use of energy from renewable sources in developing countries with high levels of economic inequality.

Keywords Energy consumption · Renewable sources · Economic inequality · Panel data

Y. Revtiuk (✉)
Poznan University of Technology, 2 J. Rychlewskiego Str., 60965 Poznan, Poland
e-mail: yevhen.revtiuk@put.poznan.pl

V. Koval
Izmail State University of Humanities, 12 Repina Str., Izmail 68610, Ukraine

I. W. E. Arsawan
Politeknik Negeri Bali, Bukit Jimbaran, Kuta Selatan 80364, Bali, Indonesia

© The Author(s), under exclusive license to Springer Nature Switzerland AG 2024
V. Koval (ed.), *Renewables in the Circular Economy and Business*,
SpringerBriefs in Applied Sciences and Technology,
https://doi.org/10.1007/978-3-031-72174-8_1

1 Introduction

Developed countries have demonstrated the growing dominance of renewable energy sources in the energy balance. Green energy accounts for 30% of a country's energy needs. The share of renewable energy sources in EU countries varies by 40–50% of the total production volume, with a decrease in electricity production considering fossil fuels [1]. Technological development contributes to an increased share of renewable energy in the energy balance and allows countries to exploit more stable natural resources. Globally, the share of renewable energy sources in the energy balance also increases; for example, in 2022, it amounted to 30% (+ 1.5 points), 10 points higher than that in 2010 [2].

The transformation of energy systems and the development of renewable energy sources involve a change in political and market conditions to improve the investment climate for the use of environmentally safe technologies in energy transition, which will help achieve sustainable development goals [3]. Improvements in renewable technologies and increased supply chain efficiency have led to lower introduction costs in the new markets.

Investments in renewable energy, which accounts for about 58% of clean energy in 2022 [4], have also increased over the past decade. Despite the positive dynamics, global regions demonstrate disproportionate levels of investment in renewable energy: it is high in developed countries (the European Union, the USA, and China) and much lower in developing economies. Despite the recent increase in investments in India and Brazil, the total volume of investments in renewables in most developing countries remained constant in 2015 [3].

Therefore, it is important to study the macroeconomic factors that determine the production and consumption of renewable energy. Studies often use panel data for analysis [5–8], including various macroeconomic variables (GDP, economic inequality, and investment) that affect the adoption and diffusion of renewable energy technology. To complement previous research, this study analyzes macroeconomic indicators and their association with renewable energy consumption using a cross-country panel data approach for 2000–2020.

2 Theoretical Background

Economic inequality refers to the unequal distribution of economic resources among members of society. Household income and consumption expenditure measure economic inequality [9]. Household income includes all income received by household members during one year minus income tax and outgoing transfers. Household consumption is the net income minus savings. The Gini coefficient is the most common indicator for assessing household income inequality. It characterizes the deviation of the Lorentz curve (the curve that shows the share of the country's total income received by a particular share of the country's households) from the ideal curve when household income is evenly distributed [10].

Researchers have primarily analyzed the relationship between economic inequality and energy use from the perspective of access to energy by economically deprived social groups [11]. Studies suggest that reducing the level of economic inequality leads to a reduction in the level of poverty and, as a result, improves groups' access to energy [12] and contributes to the growth of energy consumption. A similar mechanism can be at work in the case of energy consumption from renewable sources, as increased access to energy leads to an increase in the energy demand and, as a result, stimulates the production of energy, including that from renewable sources. On the other hand, in countries with high levels of economic inequality, policymakers introduce government policies to simultaneously reduce the level of economic inequality and environmental degradation, including stimulating the use of renewable energy sources [13]. Thus, one may hypothesize that:

H.1.1 Economic Inequality is Negatively Associated with Renewable Energy Consumption

However, the relationship between economic inequality and economic growth remains ambiguous. Political economy literature suggests that economic inequality negatively affects economic growth through social instability [14], financial market imperfections and indivisibilities in investment in human capital [15], more redistribution through the political process, and engagement of disadvantaged groups in crime, riots, and other disruptive activities [16]. On the other hand, economic inequality stimulates capital accumulation and facilitates entrepreneurs' access to financial resources [17] and, as a result, promotes long- and medium-term economic growth [18].

The "inverted-U" hypothesis by Kuznets [19] addresses this ambiguity. According to the inverted-U approach, economic growth leads to an increase in economic inequality within the country and then to a decrease, which is empirically confirmed by research [20–22]. In this case, the positive effect of economic inequality on access to financial resources can lead to an increase in the production of energy from renewable sources and, as a result, an increase in renewable energy consumption. For these reasons, the following competing hypothesis is formulated:

H1.2. Economic Inequality is Positively Associated with Renewable Energy Consumption

Debates on the relationship between the level of economic development and the use of renewable energy sources are ongoing. Most researchers point to a direct impact of energy production from renewable sources on economic performance [23–25]. At the same time, this effect depends on many factors, such as technological development, energy policies, or income level. On the other hand, the impact of economic growth on renewable energy consumption has remained out of the academic focus, with few exceptions. Omri and Nguyen, for example, suggested a mutually reinforcing effect between economic development and energy production from renewable sources [26], which they empirically confirmed for high-income countries. Another study [7] demonstrated a correlation between the level of GDP and renewable energy consumption in EU countries.

At the individual level, a higher income level allows one to consume more expensive energy from renewable sources [27], including due to the reinforcement of pro-environmental behavior [28, 29]. These effects are aggregated; therefore, in high-income countries, individuals are more motivated to consume energy from renewable sources. Based on these considerations, the following hypothesis is formulated:

H2. *The Level of Economic Development is Positively Associated with Renewable Energy Consumption*

Access to investment creates additional incentives to implement more expensive and, as a rule, less profitable projects in renewable energy production. Therefore, financial development positively affects the share of energy consumption from renewable sources [5, 8]. In combination with a developed stock market, foreign direct investments can play an important role and positively impact renewable energy consumption [30]. On the other hand, analyzing the structure of domestic investments, the relationship between the amount of gross capital formation and renewable energy consumption needs to be more straightforward and sufficiently researched. The availability of cheap credit has a positive effect on the business environment and creates prerequisites for the growth of production, including producing energy from renewable sources. At the same time, taking into account the lower profitability of projects related to the production of energy from renewable sources and their dependence on the government's stimulating policy, the growth of the investment attractiveness of the economy will, first of all, stimulate producers to invest in industries with higher profitability. Therefore, less profitable projects in the field of renewable energy will be financed after the exhaustion of other more attractive investment opportunities. Guided by the above considerations, a hypothesis regarding the relationship between the share of domestic investments and renewable energy consumption is proposed:

H3. *A High Level of Investment in the Demand Structure is Negatively Associated with Renewable Energy Consumption*

The structure of the economy affects energy consumption [6]. While low-developed agricultural countries consume a small amount of energy, the transition to the industrial phase, accompanied by the development of industrial enterprises, leads to an increase in energy consumption. At the same time, big businesses have enough resources to lobby for governmental policies aimed at reducing the cost of energy and creating obstacles in the field of more costly renewable energy policies [31].

This study includes data on countries with different political regimes, different historical and cultural experiences, and different tax systems, which, in turn, affected the processes of capital formation. It was thus necessary to re-examine the role of macroeconomic indicators, including the proportion of large, medium, and small businesses in generating added value, to evaluate the concentration of production. Instead, we argue that a share of employers among the employed population is a more objective indicator. This share will be lower in countries with a high concentration of production capital. The following hypothesis was suggested:

H4. ***The Share of Employers is Negatively Associated with the Use of Renewable Energy Sources***

3 Methodology

Renewable energy consumption was operationalized as a share of final energy consumption. The latter encompasses energy consumption derived from the following sources: hydro, solid biofuels, wind, solar, liquid biofuels, biogas, geothermal, marine, and waste. The share of primary energy consumption from renewable sources is shown in Fig. 1.

To operationalize the level of economic inequality, the Gini indicator is used, which characterizes the level of income inequality of households from $0 =$ all households received the same income to $1 =$ all income received by one household. These data come from the SWIID database [33].

We operationalized the state and structure of the economy using GDP per capita and the share of employers among employees. In addition, economic activity in the country is operationalized as the gross capital formation as a percentage of GDP. This indicator characterizes the share of domestic investments in the economy. The level of income in the country is controlled according to the World Bank classification (high income $= 4$, upper middle income $= 3$, lower middle income $= 2$, low income $= 1$, not classified $= 0$ (was not included in the model)). These data come from the World Bank Open Data [34] for the period from 1990 to 2020 for 159 countries (the list of countries is provided in Annex 1).

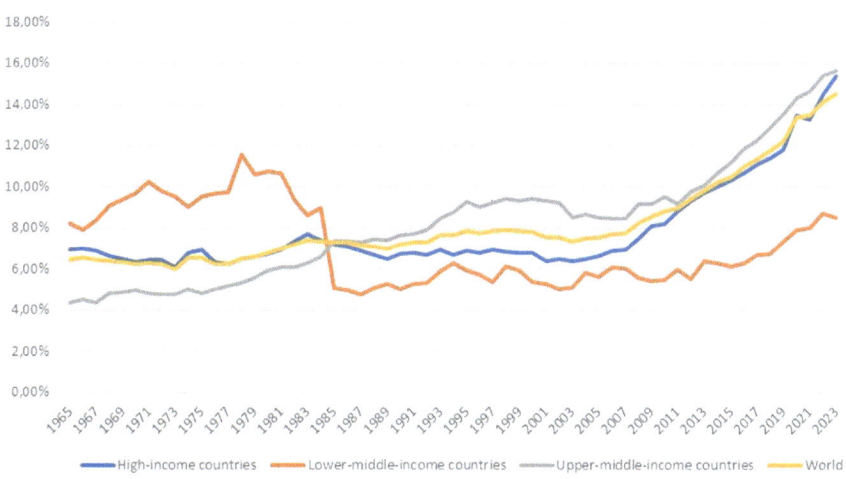

Fig. 1 Share of primary energy consumption from renewable sources [32]

Table 1 Descriptive statistics

Variable	Description	Mean	Std. Dev.	Min	Max
Renewable energy consumption	% of total energy consumption	32.433	28.904	0	97.88
Gini	%	38.696	8.659	18.5	65.4
Gross capital formation	% of GDP	23.27	7.795	− 3.946	70.335
GDP per capita	Current 1000 US$	12	17.578	0.06	123.679
Employers	% of total employment	3.428	2.485	0.009	17.881
Country level of income		2.891	1.021	0	4

Table 1 provides descriptive statistics for the indicators used in the study. A series of two-level models was employed for the analysis of the obtained longitudinal data. The shortest series had $n = 1$ observations, the longest had $n = 30$, and the average duration of observations was $n = 22.7$.

As Table 1 suggests, the share of renewable energy consumption varies from 0% in hydrocarbon-rich countries (Bahrain, Kuwait, Oman, Turkmenistan) and countries with low energy demand (Malta, Timor-Leste), to about 97% in developing countries, where relatively small energy demand is covered by the production of energy by hydroelectric plants (Democratic Republic of the Congo, Uganda), with an average share of 32%. The Gini index ranges from 18.5% (Slovak Republic in 1991) to 65.4% (Namibia in 1993–1994), averaging 39%. GDP per capita ranges from USD 0.06 thousand for Azerbaijan in 1992 to USD 123.679 thousand for Luxembourg in 2014, with an average value of USD 12 thousand.

4   Empirical Results

A series of multilevel regression models were implemented to test our hypotheses. The results are presented in Table 2. A hierarchical approach is used to construct the model. The model was tested by sequentially adding independent variables: Gini and GDP per capita (Model 1), gross capital formation (Model 2), employers (Model 3), country level of income (Model 4), interaction Gini * GDP per capita (Model 5), interaction Gini * Gross capital formation (Model 6). Verification using the likelihood ratio test demonstrated that the corresponding lower-order models were nested within higher-order models. Simultaneously, changes in the Intraclass Correlation Coefficient (ICC) are monitored. A decrease in the ICC value as an independent variable was added to the models, indicating increased accuracy. Table 2 provides the results of the model calculations using robust standard errors.

Table 2 Results of multilevel analysis, 1990–2020

	0	1	2	3	4	5	6
Gini		0.5759* (0.2599)	0.5563* (0.2546)	0.6530** (0.2455)	0.6168* (0.2409)	0.6548** (0.2369)	0.8988*** (0.2547)
GDP per capita		0.1057** (0.0345)	0.1112** (0.0359)	0.1060** (0.0333)	0.1103** (0.0336)	0.7476** (0.2619)	0.1084** (0.0333)
Gross capital formation			− 0.1937*** (0.0374)	− 0.1920*** (0.0367)	− 0.1907*** (0.0366)	− 0.1920*** (0.0357)	0.2715 (0.1987)
Employers				− 1.4861** (0.4579)	− 1.4412** (0.4483)	− 1.4383*** (0.4360)	− 1.4560** (0.4444)
Income							
Low income					60.3880*** (4.5367)	59.1832*** (4.4974)	59.8862*** (4.6191)
Lower middle income					31.7704*** (3.5512)	30.5594*** (3.5007)	31.7048*** (3.5395)
Upper middle income					8.3551** (2.9506)	7.7887** (2.8342)	8.3019** (2.9433)
High income					0.6548 (3.3501)	0.3093 (3.2341)	0.7471 (3.3053)
Gini * GDP per capita						− 0.0200** (0.0074)	
Gini * Gross capital formation							− 0.0121* (0.0053)
Constant	34.3413*** (2.3564)	10.6283 (10.4456)	16.0437 (10.6127)	17.0933 (10.6214)	0.0156 (11.1077)	− 0.6098 (10.7503)	− 10.7303 (11.3847)
Constant	− 17.9006 (15.7784)	− 5.6446*** (1.1792)	− 5.2859*** (0.5517)	− 5.3346*** (0.5828)	− 5.0893*** (0.3440)	− 5.4665*** (0.6944)	− 5.0987*** (0.3579)

(continued)

Table 2 (continued)

	0	1	2	3	4	5	6
Constant	3.3874*** (0.0412)	3.3285*** (0.0746)	3.2926*** (0.0745)	3.2397*** (0.0786)	2.7367*** (0.2034)	2.8389*** (0.1594)	2.7481*** (0.2019)
Constant	1.6698*** (0.0612)	1.6436*** (0.0651)	1.6239*** (0.0655)	1.6037*** (0.0625)	1.6032*** (0.0627)	1.5913*** (0.0634)	1.5988*** (0.0625)
AIC	23,303.6781	23,118.3397	22,983.8969	22,830.1118	22,730.8081	22,639.0473	22,703.3276
BIC	23,328.4384	23,155.4801	23,027.2275	22,879.6324	22,798.8990	22,713.3282	22,777.6085
N	3605.0000	3605.0000	3605.0000	3605.0000	3605.0000	3605.0000	3605.0000

Note Standard errors in parentheses; *$p < 0.05$, **$p < 0.01$, ***$p < 0.001$

4.1 The Unconditional Random Intercept Model

First, an empty two-level model (Model 0) without predictors was constructed (Table 2) and calculated the Intraclass Correlation Coefficient (ICC) [35], which is ICC = 0.96 and indicates a high level of within-cluster homogeneity [36], meaning that 0.04% of the variance in renewable energy consumption can be attributed to within-country differences. The calculated design effect (DEFF) of the sample is DEFF = 22.02 (DEFF > 2), which indicates that the use of multilevel models is preferable to traditional regression [37].

Figure 2 (left-hand side) shows country (two-level) residuals for the unconditional random intercept model, which demonstrates the existence of between-country differences across countries. The right-hand side of Fig. 2 shows the countries for which the confidence interval crosses the line at zero (representing the mean renewable energy consumption value across all countries). For other countries, it can be argued that residuals differ significantly from the average at the 5% level.

4.1.1 The Two-Level Random Intercept Models

Several two-level random intercept models (Models 1–4 in Table 2) were tested due to the addition of independent and control variables. At each stage of the analysis, the fit of the models was evaluated using the likelihood ratio test.

Model 1 estimates the effect of the Gini coefficient and GDP per capita on renewable energy consumption without controlling for any other characteristics. The coefficients of Gini are positive and statistically significant ($p < 0.05$), which confirms hypothesis 1.2 and refutes hypothesis 1.1.

The coefficient of GDP per capita is positive and statistically significant ($p < 0.01$), which in turn confirms hypothesis 2 that the level of economic development is positively associated with renewable energy consumption.

The Gross capital formation indicator was added to Model 2 and turned a negative and statistically significant ($p < 0.001$) regressor of renewable energy consumption. Thus, hypothesis 3 is confirmed. In the next step, Model 3 contains the employment indicator. Similarly to the case of the previous regressor, a share of employers is negatively and significantly ($p < 0.01$) associated with renewable energy consumption, which confirms hypothesis 4. The country's income level does not change the nature of the relationship between the repressors described above, nor their statistical significance. At the same time, the countries belonging to low-income, lower-middle-income, and upper-middle-income groups have a statistically significant association with the share of renewable energy consumption.

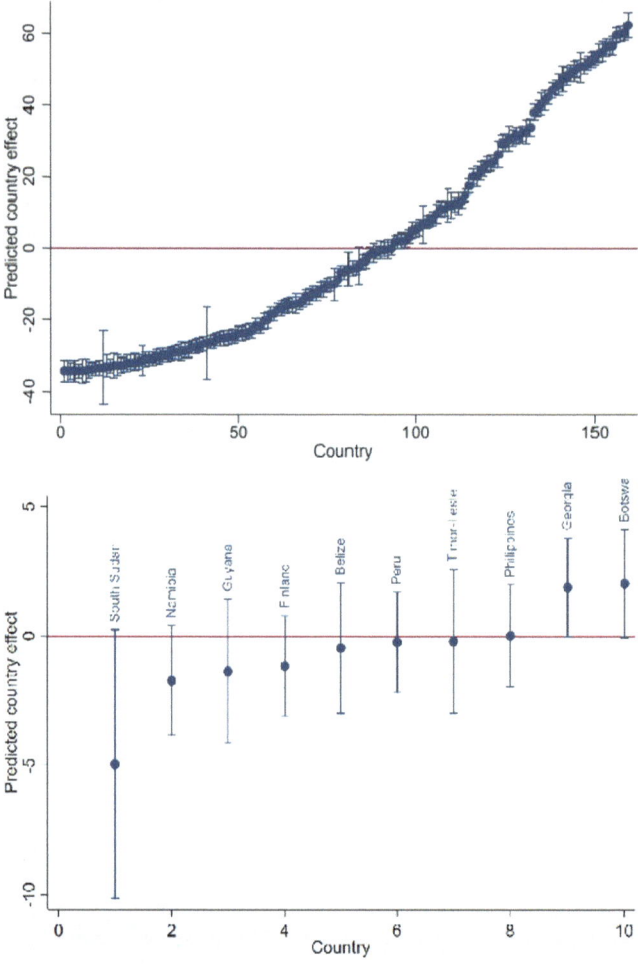

Fig. 2 Country residuals for unconditional random intercept model, 1990–2020

4.1.2 The Two-Level Random Intercept Models with Interaction

To assess the mutual association of Gini and GDP per capita, and Gini and gross capital formation with the share of renewable energy consumption, the interaction between Gini and GDP per capita (Model 5) and the interaction between Gini and gross capital formation (Model 6) were included in the models. In Model 5 (Table 2), the independent variables remain significant, and adding the interaction term does not change the nature of the association. The interaction between Gini and GDP per capita is negatively associated with renewable energy consumption, and this relationship is statistically significant ($p < 0.01$). Therefore, GDP per capita has a robust negative effect on the association of Gini with renewable energy consumption. For countries with a lower GDP per capita, renewable energy consumption is more inelastic to economic inequality and vice versa.

The association between the interaction between Gini and gross capital formation and renewable energy consumption is also negative and statistically significant ($p <$ 0.01). At the same time, the inclusion of this interaction term in the model changes the nature of the association of the gross capital formation indicator from negative to positive, although this association is statistically insignificant. It is therefore not possible to state with certainty that countries with a higher level of economic inequality and a higher indicator of gross capital formation will experience a lower level of renewable energy consumption [38, 39].

5   Conclusions

The above research suggests that economic inequality is positively associated with renewable energy consumption (i.e., a 1% increase in economic inequality is associated with a 0.58% increase in renewable energy consumption). Thus, countries with a higher concentration of income in the affluent population stratum consume more energy from renewable sources. This result can be explained by the following interrelated factors: (1) wealthier strata consume more energy, (2) a higher level of income allows them to buy more expensive renewable energy, and (3) demonstrate a pro-environmental position. The positive association between GDP per capita and renewable energy consumption serves as additional confirmation of our assumption. Conversely, the findings revealed a negative mutual association between the Gini coefficient, GDP per capita, and renewable energy consumption. Thus, in countries with lower GDP per capita, the positive association between economic inequality and renewable energy consumption is felt to a lesser extent, and vice versa. That is, wealthy population strata in countries with a lower GDP per capita will have fewer resources than rich strata in countries with a higher GDP per capita and, therefore, will be less inclined to consume more expensive renewable energy.

The analysis also revealed a negative association between gross capital formation and renewable energy consumption. This partially contradicts the consensus among scientists regarding the positive relationship between access to investment and the degree of development of the financial system and renewable energy consumption. One explanation could be that big businesses today tend to finance more profitable projects in the economy's traditional branches with a higher profitability level than projects related to energy production from renewable sources. In addition, big businesses are interested in accessing cheaper energy sources and, therefore, will hinder or at least refrain from lobbying for governmental policies to increase the share of renewable energy in the balance of energy consumption. The fact that the share of employers among the employed population is negatively associated with renewable energy consumption can serve as additional evidence for this assumption, considering that for countries with a larger share of businesses in the production structure, the indicator of the share of employers among the employed population is higher.

Annex 1. List of the Countries and Number of Observations Included in Models

Albania (24), Algeria (21), Angola (17), Argentina (29), Armenia (30), Australia (28), Austria (30), Azerbaijan (18), Bahamas (13), Bahrain (1), Bangladesh (26), Barbados (27), Belarus (28), Belgium (30), Belize (17), Benin (14), Bhutan (15), Bolivia (29), Bosnia and Herzegovina (15), Botswana (25), Brazil (29), Bulgaria (30), Burkina Faso (21), Burundi (22), Cambodia (16), Cameroon (19), Canada (30), Central African Republic (17), Chad (9), Chile (27), China (30), Colombia (30), Comoros (11), Dem Rep Congo (9), Rep Congo (7), Costa Rica (30), Cote d'Ivoire (25), Croatia (25), Cyprus (29), Denmark (30), Djibouti (5), Dominican Republic (29), Ecuador (26) Egypt (27), El Salvador (29), Equatorial Guinea (1), Estonia (26), Eswatini (26), Ethiopia (21), Fiji (23), Finland (29), France (29), Gabon (13), Gambia (25), Georgia (30), Germany (29), Ghana (26), Greece (29), Guatemala (24), Guinea (22), Guinea-Bissau (18), Guyana (14), Haiti (12), Honduras (29), Hong Kong SAR (26), Hungary (29), Iceland (26), India (30), Indonesia (30), Iran (26), Iraq (9), Ireland (30), Israel (24), Italy (30), Jamaica (27), Japan (28), Jordan (27), Kazakhstan (29), Kenya (25), Rep. Korea (30), Kuwait (9), Kyrgyz Republic (28), Lao PDR (17), Latvia (25), Lebanon (18), Lesotho (11), Libya (6), Lithuania (25), Luxembourg (29), Madagascar (22), Malaysia (29), Mali (27), Malta (21), Mauritania (24), Mauritius (27), Mexico (28), Moldova (26), Mongolia (24), Montenegro (15), Morocco (24), Namibia (24), Nepal (20), the Netherlands (29), New Zealand (28), Nicaragua (22), Niger (23), North Macedonia (26), Norway (30), Oman (11), Pakistan (28), Panama (27), Papua New Guinea (9), Paraguay (30), Peru (29), the Philippines (28), Poland (26), Portugal (29), Puerto Rico (26), Qatar (20), Romania (29), Russian Federation (30), Rwanda (26), Samoa (4), Saudi Arabia (12), Senegal (21), Serbia (23), Sierra Leone (28), Singapore (30), Slovak Republic (29), Slovenia (25), Solomon Islands (8), South Africa (27), South Sudan (4), Spain (29), Sri Lanka (21), Sudan (24), Suriname (5), Sweden (30), Switzerland (29), Tajikistan (23), Tanzania (27), Thailand (29), Timor-Leste (14), Togo (11), Tonga (25), Tunisia (25), Turkmenistan (12), Uganda (26), Ukraine (29), the United Arab Emirates (11), the UK (30), the USA (30), Uruguay (30), Uzbekistan (12), Vanuatu (5), Venezuela (24), Viet Nam (26), Zambia (6), Zimbabwe (25).

References

1. EEA, Share of energy consumption from renewable sources in Europe (2024). https://www.eea.europa.eu/en/analysis/indicators/share-of-energy-consumption-from
2. Enerdata (2022). https://energystats.enerdata.net
3. IRENA, *Annual Renewable Power Must Triple Until 2030* (2023). https://www.irena.org/News/pressreleases/2023/Jun/Annual-Renewable-Power-Must-Triple-by-2030
4. IEA, *Overview and Key Findings—World Energy Investment 2022—Analysis* (IEA, 2022). https://www.iea.org/reports/world-energy-investment-2022/overview-and-key-findings
5. S.G. Anton, A.E. Afloarei Nucu, The effect of financial development on renewable energy consumption A panel data approach. Renew. Energy **147**, 330–338 (2020). https://doi.org/10.1016/j.renene.2019.09.005
6. M. Can, Z. Ahmed, Towards sustainable development in the European Union countries: does economic complexity affect renewable and non-renewable energy consumption? Sustain. Dev. **31**, 439–451 (2023). https://doi.org/10.1002/sd.2402
7. S. Ntanos, M. Skordoulis, G. Kyriakopoulos, G. Arabatzis, M. Chalikias, S. Galatsidas, A. Batzios, A. Katsarou, Renewable energy and economic growth: evidence from European countries. Sustainability **10** (2018). https://doi.org/10.3390/su10082626
8. L. Wu, D.C. Broadstock, Does economic, financial and institutional development matter for renewable energy consumption? Evidence from emerging economies. Int. J. Econ. Policy Emerg. Econ. **8**, 20–39 (2015). https://doi.org/10.1504/IJEPEE.2015.068246
9. S.P. Jenkins, P. Van Kerm, The measurement of economic inequality, in *Oxford Handbook on Economic Inequality* (Oxford University Press, 2011), pp. 40–67
10. A. Sen, *On Economic Inequality* (Oxford University Press, 1997)
11. R. Galvin, Inequality and energy: how extremes of wealth and poverty in high income countries affect CO_2 emissions and access to energy, in *Inequality and Energy: How Extremes of Wealth and Poverty in High Income Countries Affect CO_2 Emissions and Access to Energy* (2019). https://doi.org/10.1016/C2018-0-02082-4
12. M. Aghaei, C.-Y.C. Lin Lawell, Energy, economic growth, inequality, and poverty in Iran. Singapore Econ. Rev. **67**, 733–754 (2022). https://doi.org/10.1142/S0217590820500198
13. U. Uzar, Is income inequality a driver for renewable energy consumption? J. Clean. Prod. **255** (2020). https://doi.org/10.1016/j.jclepro.2020.120287
14. A. Alesina, R. Perotti, Income distribution, political instability, and investment. Eur. Econ. Rev. **40**, 1203–1228 (1996). https://doi.org/10.1016/0014-2921(95)00030-5
15. O. Galor, J. Zeira, Income distribution and macroeconomics. Rev. Econ. Stud. **60**, 35–52 (1993). https://doi.org/10.2307/2297811
16. R.J. Barro, Inequality and growth in a panel of countries. J. Econ. Growth **5**, 5–32 (2000). https://doi.org/10.1023/A:1009850119329
17. P.C. Patel, J.P. Doh, S. Bagchi, Can entrepreneurial initiative blunt the economic inequality-growth curse? Evidence from 92 countries. Bus. Soc. **60**, 496–541 (2021). https://doi.org/10.1177/0007650318797103
18. C.M. Menyelim, A.A. Babajide, A.E. Omankhanlen, B.I. Ehikioya, Financial inclusion, income inequality and sustainable economic growth in Sub-Saharan African countries. Sustainability **13**, 1–15 (2021). https://doi.org/10.3390/su13041780
19. S. Kuznets, Economic growth and income inequality. Am. Econ. Rev. **45**, 1–28 (1955)
20. J.G. Brida, E.J.S. Carrera, V. Segarra, Clustering and regime dynamics for economic growth and income inequality. Struct. Chang. Econ. Dyn. **52**, 99–108 (2020). https://doi.org/10.1016/j.strueco.2019.09.010
21. M.Á. Caraballo, C. Dabús, F. Delbianco, Income inequality and economic growth revisited. A note. J. Int. Dev. **29**, 1025–1029 (2017). https://doi.org/10.1002/jid.3300
22. F. Fawaz, M. Rahnama, V.J. Valcarcel, A refinement of the relationship between economic growth and income inequality. Appl. Econ. **46**, 3351–3361 (2014). https://doi.org/10.1080/00036846.2014.929624

23. C. Chen, M. Pinar, T. Stengos, Renewable energy consumption and economic growth nexus: evidence from a threshold model. Energy Policy **139** (2020). https://doi.org/10.1016/j.enpol. 2020.111295
24. M. Shahbaz, C. Raghutla, K.R. Chittedi, Z. Jiao, X.V. Vo, The effect of renewable energy consumption on economic growth: evidence from the renewable energy country attractive index. Energy **207** (2020). https://doi.org/10.1016/j.energy.2020.118162
25. Q. Wang, Z. Dong, R. Li, Revisiting renewable energy and economic growth—does trade openness a matter? Environ. Sci. Pollut. Res. **30**, 31727–31740 (2023). https://doi.org/10.1007/s11356-022-24358-x
26. A. Omri, D.K. Nguyen, On the determinants of renewable energy consumption: international evidence. Energy **72**, 554–560 (2014). https://doi.org/10.1016/j.energy.2014.05.081
27. R. Sharma, S.S. Rajpurohit, Nexus between income inequality and consumption of renewable energy in India: a nonlinear examination. Econ. Chang. Restruct. **55**, 2337–2358 (2022). https://doi.org/10.1007/s10644-022-09389-1
28. S. Du, G. Cao, Y. Huang, The effect of income satisfaction on the relationship between income class and pro-environment behavior. Appl. Econ. Lett. (2022). https://doi.org/10.1080/13504851.2022.2125491
29. S. Otto, A. Neaman, B. Richards, A. Marió, Explaining the ambiguous relations between income, environmental knowledge, and environmentally significant behavior. Soc. Nat. Resour. **29**, 628–632 (2016). https://doi.org/10.1080/08941920.2015.1037410
30. A.M. Kutan, S.R. Paramati, M. Ummalla, A. Zakari, Financing renewable energy projects in major emerging market economies: evidence in the perspective of sustainable economic development. Emerg. Mark. Financ. Trade **54**, 1761–1777 (2018). https://doi.org/10.1080/1540496X.2017.1363036
31. I. Cadoret, F. Padovano, The political drivers of renewable energies policies. Energy Econ. **56**, 261–269 (2016). https://doi.org/10.1016/j.eneco.2016.03.003
32. H. Ritchie, M. Roser, P. Rosado, Renewable energy sources are growing quickly and will play a vital role in tackling climate change. Our World Data (2024)
33. F. Solt, *The Standardized World Income Inequality Database, Versions 8–9* (2023).https://doi.org/10.7910/DVN/LM4OWF
34. World Bank Open Data, Data (2024). https://data.worldbank.org/. Accessed 4 July 19
35. J.J. Hox, M. Moerbeek, R. van de Schoot, *Multilevel Analysis: Techniques and Applications* (Routledge, 2017).https://doi.org/10.4324/9781315650982
36. I.G. Kreft, J. de Leeuw, *Introducing Multilevel Modeling*, (SAGE Publications, 1998)
37. J.L. Peugh, A practical guide to multilevel modeling. J. Sch. Psychol. **48**(1), 85–112 (2010). https://doi.org/10.1016/j.jsp.2009.09.002
38. V. Koval, Y. Sribna, S. Kaczmarzewski, A. Shapovalova, V. Stupnytskyi, Regulatory policy of renewable energy sources in the European national economies. Polityka Energetyczna Energy Policy J. **24**(3), 61–78 (2021). https://doi.org/10.33223/epj/141990
39. S.J. Ergun, M.F. Rivas, Does higher income lead to more renewable energy consumption? Evidence from emerging-Asian countries. Heliyon **9**(1), e13049 (2023). https://doi.org/10.1016/j.heliyon.2023.e13049

Strategic Development and Implementation of Sustainable Energy Initiatives

Lyudmila Poppⓘ**, Kairat Imanbekov**ⓘ**, Gulzada Mukhamediyeva**ⓘ**, Faiza Bokizhanova**ⓘ**, and Olga Kobzareva**ⓘ

Abstract This study outlines key strategies to create and implement long-term energy plans, as evidenced by 15 case studies. Renewable energy and efficient companies such as Tesla, Siemens, and Vestas have presented higher value added by industry, but they emit less carbon. For instance, Tesla achieved a 3.5% value added by industry and a 50% reduction in carbon emissions with a 50% renewable energy share commitment but obtained 4%, that is, a 0% value added by industry with a 60% renewable energy share. There is thus a requirement for harmonization and for the review of these coordination goals such as how policies are aligned and incentives offered as well as the incorporation of topics like sustainable energy policies into broader policy frameworks. Pursuant to the econometric model, a 10% boost in renewable energy use will arise in a 1.2% gain in industrial value added and a 5% decrease in carbon emissions. These findings are supported by case studies of companies that emphasize the importance of a complete plan for sustainable energy development.

Keywords Renewable energy consumption · Energy efficiency · Economic growth · Carbon emissions · Policy coordination

1 Introduction

Sustainable energy development, therefore, provides energy to the current society in a manner that a society a hundred years from now shall not be in a position to provide for itself. It involves the use of energy sources, such as solar, wind, and hydropower, and embracing efficiency techniques in using energy to reduce greenhouse emissions

L. Popp (✉)
Toraighyrov University, 64 Lomova Str., 140008 Pavlodar, Republic of Kazakhstan
e-mail: inforws212@gmail.com

K. Imanbekov · G. Mukhamediyeva · F. Bokizhanova · O. Kobzareva
Q University, 125/185 Bayzakova Str., 050000 Almaty, Republic of Kazakhstan

© The Author(s), under exclusive license to Springer Nature Switzerland AG 2024
V. Koval (ed.), *Renewables in the Circular Economy and Business*,
SpringerBriefs in Applied Sciences and Technology,
https://doi.org/10.1007/978-3-031-72174-8_2

and impact [1]. The importance of sustainable energy development can be attributed to the fact that it affects two of the three dimensions of sustainability [2].

Adjustment in climate can be made if transition is to be made to other forms of energy after utilizing existing sources of energy. Therefore, the generation of sustainable energy is beneficial in a number of ways economically [3].

Self-sustaining niche markets can also be created in renewable energy (RE), technological advancement of new clean technologies, and enhanced energy security through diversification and reduction in the use of imported fuel. Second, expenditure on sustainable energy can lead to long-term gains; that is, expenditure on renewable energy forms is less than that of fossil energy.

Thus, the development of sustainable energy requires a complex approach, which will solve the problems of energy security, climate changes, and economic growth [4]. It should also look at the social and economic impacts of change toward renewable energy, particularly on sensitive companies.

An efficient policy approach must integrate the energy security, climate change, and economic development sectors to develop a more sustainable and better-balanced energy system that is effective and detrimental to the environment [5, 6]. Although people have recently realized the value of investing in sustainable energy futures, most parties at the national or organizational level need to implement sustainable energy policies efficiently. This becomes a challenging task to meet societal needs for environmental sustainability and economic development.

The aim is to present the conceptual model for synthesizing and implementing the strategies for sustainable development of energy resources. The framework is also intended to help policymakers, businesses, and other stakeholders adapt to sustainable energy more efficiently, as the structured approach is intended to present strategies for following this course. The objectives of this study can be summarized in two goals.

First, it aims to enable readers to fully appreciate the crucial role that sustainable energy development strategies play in attaining both environmental sustainability and economic development.

Second, guidelines on how these strategies should be designed and managed should be presented. Through the framework provided, one will be better positioned to develop and implement sustainable energy practices within policy, business, and other institutions.

These elements include using renewable energy sources, energy conservation measures, policy measures, and stakeholder participation. To further expand the database of knowledge on sustainable energy development, the provisions discussed in this article outline the proposed framework. It offers pragmatic advice to decision-makers and practitioners on how a sustainable energy solution for future generations may be attained.

It aims to contribute to constructing sustainable energy systems for the future. These factors make the current study relevant to sustainable energy development and the difficulties arising during sustainable energy strategy creation and management.

This study analyzes the current sustainable energy development strategies literature and emphasizes their problems and shortcomings. The study then outlines a strategic framework for constructing and implementing sustainable energy initiatives with sub-topics including renewable energy, energy conservation, policy systems, and stakeholders' involvement plans. Some of the case studies are discussed, and after all of these, the results are identified. The conclusion discusses the main results, notes the methodology's usefulness, and suggests further research and practical recommendations.

2   Literature Review

The task of creating and implementing the energy development plan at the same time is a complex process and a prerequisite for understanding the impact of various factors. This literature review focuses on the recent research relating to specific issues, including natural resources, economic growth, renewable energy, and policies.

For instance, Bekun et al. are confined merely to analyze the correlation between CO_2 emission, energy resource rent, and renewable/non-renewable energy in 16 EU countries. They pointed out this transition as essential in their research to decrease the level of carbon dioxide emissions and, in the process, embrace environmental conservation [2]. This insight is of great importance when developing strategies for energy investments in renewable energy fields.

Candra et al. [7] studied how the adoption of renewable energy helps in improving on the economic sustainability and at the same time decreases on the emission of greenhouse gases. This evidence subscribes to the call for incorporating renewable energy in the country's development policies.

Canh and Thong describe how financialization links to natural resource rents and present a cross-country analysis of the case [8]. Thus, they argue that financialization can affect the distribution of natural resource rents, which is relevant to sustainable energy outcomes. This perspective is useful in the fact that it provides a view of the economic layers that define energy resource management.

Jia, Fan, and Xia explore how the use of renewable energy impacts the economic development of Belt and Road countries [9]. On their part, they have noted that renewable energy affects growth and has an energy consumption increment, which implies the importance of renewable energy in promoting sustainable economic growth. This is why it is imperative that policies that bear the potential of enabling the integration of renewable energy be implemented.

Concerning the Polish national courts' struggles with rights-based smog cases, Karpus [4] reflects on this. As this study shows, the legal and regulation framework in the context of the analysis identifies the key hurdles in implementing environmental policies, which is essential for understanding the issues in the progress of sustainable energy. The policies regulating the utilization of energy need to be well responded to through legal systems.

The study by Lu and Wu [10] demonstrates that selective FDI can promote the green competitiveness within industries; in other words, inviting green investments is a possibility for furthering sustainable energy. This corresponds to the requirement of strategic economic policies aimed at supporting green investments. Murshed examines if the strategies of trade liberalization are consistent with the shift to renewable sources in low- and middle-income LMICs [5]. The study in consideration shows that external policies must be restructured in order to promote the growth of renewable energy using the instrumental variable analysis. This insight is noble for coordinating trade and energy policies for development to enhance the public's welfare.

Ren et al. propose a wavelet quantile on quantile analysis to investigate the co-movement between the carbon market and the green bonds market [11]. They also identify the ways by which the financial aspects can help in the achievement of change toward a green economy. It is crucial to comprehend these flows in order to devise tactics, which may harness financial markets for the purposes of financially sustaining sustainability in the energy sector [12].

Specifically, regarding reusability, Xiao et al. are interested in the process of cathode regeneration for used lithium-ion batteries. Therefore, their work is more sustainability-centered concerning resource recycling [13]. Other studies echo the need to create new strategies for waste from energy technology and concerning the circular economy and sustainable waste management in particular [14].

Muller presents a case of hydrothermal liquefaction of spent coffee grounds and their upgradation by biocatalytic conversion for biofuel generation which embodies the circular economy utilization of waste resources [15]. The findings presented in this research illustrate the possibility of sustainability by turning waste streams into valuable commodities and resources. Bhola carries out a techno-economic and environmental feasibility study on the use of building rooftops of a campus for producing solar photovoltaic power [16]. Thus, the study affirms the possibilities and advantages of incorporating RE sources into the presently installed structure, showing the solar energy contribution to sustainable development [17, 18].

The possibilities of managing wind and solar power plant end-of-life equipment in Ukraine are explored by Trypolska et al., stressing the need for suitable approaches in the field of renewable energy [19]. Some of the prior works similar to the paper under discussion in which they discuss energy poverty and energy efficiency in emerging economies to identify obstacles and possible directions in enhancing the quality of energy supply [20]. They further explain that sustainable energy and development aid in mitigating social and economic inequities.

Dovhan et al. offer a PM textbook containing sections corresponding to sustainable energy development issues, including managing international projects, innovations, and logistics [21]. Although it is not a research book, this textbook provides essential information to comprehend project management regarding sustainable energy projects.

Kurbatova et al. describe solar energy's economic, environmental, and image gains in a university context [22]. This study shows that solar energy, the energy source, is not only seen as a portfolio of sustainable energy but as an opportunity to uplift and build the sustainability portfolio of institutions of learning. Sopronenkov et al. critically review the effects of the tax policy on businesses' prospects and economic performance [23]. Even though the subject of this study does not directly deal with sustainable energy, this paper under-emphasizes the role of policy environments in structuring economic undertakings, including sustainable energy projects.

Koval et al. [24] substantiated approaches to fiscal policy and regulation of renewable energy sources in the national European economy. This study highlights the importance of policies, especially in developing energy management practices. It also affirms the significance of combined efforts to take better strides toward sustainable energy progress. The research proposed that cooperation with partners in other countries might enhance the prospects of disseminating necessary knowledge and applying helpful technologies to support sustainable energy policy development [25].

Todorov et al. seek to identify the correlation between financial literacy and carbon footprint [26]. Increasing financial knowledge will promote better use of resources and, therefore, can help lower carbon emissions. On this basis, this study points to the fact that increasing financial knowledge can be a significant element of effective solutions to address the effects of energy in relation to minimizing carbon emissions.

Sribna et al. [27] discussed the technological implications of energy transition focusing on the role of economics in assuring the environment's safety, underscoring the importance of technology availability and sharing in improving practical power networks. The highlight of this study revolves around the subject of technology enhancement and technology transfer concerning sustainable energy.

3   Methodology

It is for this reason that, this research used a holistic approach in assessing the effect of sustainable energy development policy instruments on economic growth and sustainability of the environment. Data collection was basically through secondary sources from established sources that included World Bank [28], International Monetary Fund [29], and International Energy Agency [30]. The variables included in the dataset were renewable energy consumption, energy efficiency, value added by industry, CO_2 emissions, population growth rates, inflation rates, forest area, and annual precipitation, and the included years were 2018–2023.

The included years were 2018–2023.

The data-gathering process considers data entropy and is checked and cleaned to make the data coherent and reduce the number of outliers or missing observations. Thus, panel data models were employed to assess the impact of sustainable energy development strategies (reflected by REN and EE Indicators) on the economic-environmental performance indicators CCI and value added by industry. A qualitative aspect was included through the use of case studies of companies and countries as part of the completion of the study. These case studies were useful in giving real-life examples on how sustainable energy plans could be undertaken and the results that would be likely to be realized. The criteria for choosing the case included the type of industry, geographical location, and the success of the sustainability of energy cases. Regression tests were performed to identify the non-triviality of the links between the applied sustainable energy strategies and the studied economic-environmental variables.

The findings were used in a manner that would enable assessment of the interventions that underpin sustainable energy on economic growth and sustainable natural resource use. The last section of the study elaborated on the general limitation of the research methodology, data availability, and assumptions that were made during the analysis. The use of this methodology presented a clear framework for the interaction between sustainable energy development, economic development, and environmental sustainability, which was beneficial for policy makers, firms, and researcher.

4 Results

This issue has attracted more focus toward policy, planning, development, and management of sustainable energy strategies in the face of global challenges such as climate change, sustainable development, and skyscraper growth [31]. Sustainable energy policies and plans embrace the desirable objectives of energy security, reduced carbon intensity, and economic development by increasing renewable energy technologies and energy efficiency. A framework is further developed to reflect the impact of sustainable energy resources and their contribution to economic development and the environment. Based on this context, the proposed model combines theoretical assumptions from economics, environmental science, and energy policy to offer a systemic vision of the relationship between sustainable energy development and socioeconomic factors. The study presents a conceptual framework and procedures for developing such a model using specialized programs such as Stata for actual implementation. The model is planned to reveal to what extent the introduced advantages affect the state's economic growth and environmental sustainability by implementing further development strategies for sustainable energy resources. It considers components discussed in the paper, illustrated in Fig. 1.

This model includes the following components:

(1) Economic growth model:

- *Value_added_by_industry*—Value added by industry ratio (the percentage increase in GDP, reflecting the company's contribution to economic growth).
- $\beta 0$—the intercept term, characterizing the base equal of GDP growth when the independent variables are 0.
- $\beta 1$—the coefficient of the Renewable_energy_consumption, representing the effect of increasing renewable energy consumption on the GDP growth. A positive coefficient indicates that higher renewable energy consumption is associated with higher GDP growth, suggesting a positive impact of sustainable energy development on economic growth.
- $\beta 2$—the coefficient of Energy_efficiency, representing the impact of improved energy efficiency on GDP growth. A positive coefficient suggests

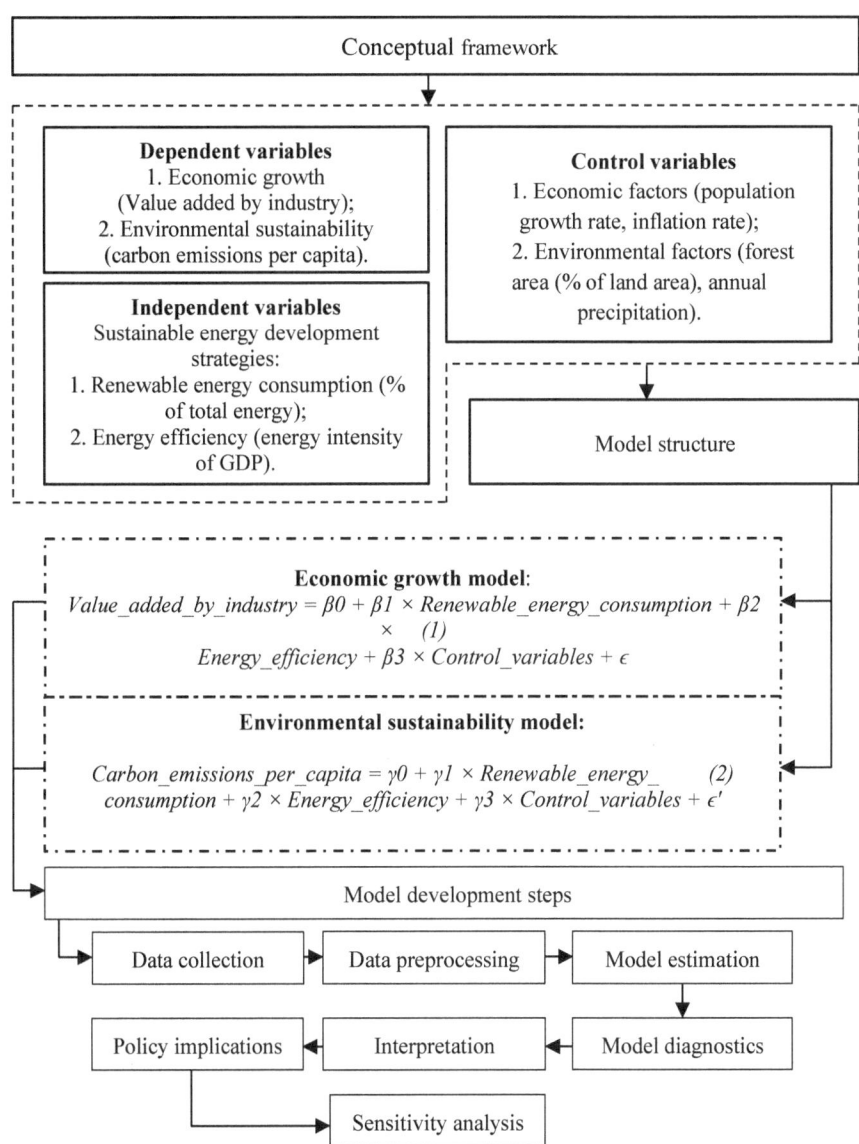

Fig. 1 Conceptual framework of the model

that higher energy efficiency is associated with higher GDP growth, indicating that energy efficiency measures can contribute to economic growth.

- $\beta 3$—the coefficient of Control_variables, representing the combined impact of other control variables (such as population growth rate and inflation rate) on GDP growth. These control variables are added in order trace other variables that have impacts on economic growth, despite the fact that they may not be captured by the model.

- ϵ—the error term, characterizing the unexplained variation in GDP growth that is not accounted for by the independent variables in the model.

(2) Environmental sustainability model:

- Carbon_emissions_per_capita—the dependent variable representing the amount of carbon emissions per capita, which is a measure of environmental sustainability.
- $\gamma 0$—the intercept term, representing the base level of carbon emissions per capita when all independent variables are zero.
- $\gamma 1$—the coefficient of Renewable_energy_consumption, representing the impact of increasing renewable energy consumption on carbon emissions per capita. A negative coefficient suggests that higher renewable energy consumption is associated with lower carbon emissions per capita, indicating a positive impact of renewable energy on environmental sustainability.
- $\gamma 2$—the coefficient of Energy_efficiency, representing the impact of improved energy efficiency on carbon emissions per capita. A negative coefficient suggests that higher energy efficiency is associated with lower carbon emissions per capita, indicating that energy efficiency measures can contribute to environmental sustainability.
- $\gamma 3$—the coefficient of Control_variables, representing the combined impact of other control variables on carbon emissions per capita.
- ϵ'—the error term, representing the unexplained variation in carbon emissions per capita that is not accounted for by the independent variables in the model.

These models assist in analyzing the connection between sustainable energy development strategies, and economic development, as well as environmental conservation. The coefficients of the independent variables show how certain degrees of renewable energy consumption and energy efficiency affect these outputs while the control variables explain confounding variables that may affect the results. From the findings on the SMG relationship with economic development and the impact of sustainable energy development strategies, the authors employ panel data model working with the data collected from fifty firms for the time frame 2018–2023 (Table 1; Fig. 2).

There is a direct positive correlation between the amount of renewable energy used and corporate energy efficiency indices, on the one hand, and the value added by industry. For instance, the USA's Tesla records a value added by industry of 3.5%, with a renewable energy consumption of 30% and achieving an efficiency of 0.05. Similar to the previous method, in this case, more irrefutable evidence comes from Siemens (Germany) and Vestas (Denmark) with contribution rates of 0.402 and 0.603, respectively, for value added by industry, which is greater than the benchmark data. At the same time, their renewable energy consumption is 40 and 60% and moderate energy efficiency with a score of 0.04 and 0.03 theirs.

Table 1 Results (fragment)

No.	Company	Country	Value added by industry (%)	Carbon emissions (MtCO$_2$)	Renewable energy consumption (%)	Energy efficiency	Population growth rate (%)	Inflation rate (%)	Forest area (% of land)	Annual precipitation (mm)
1	Tesla	USA	3.5	250	30	0.05	0.8	2.0	30	800
2	Siemens	Germany	2.0	150	40	0.04	0.2	1.5	32	900
3	Toyota	Japan	1.8	180	20	0.06	0.3	1.0	65	2000
4	Vestas	Denmark	2.5	120	60	0.03	0.1	2.5	50	1000
5	Enel	Italy	2.2	140	35	0.04	0.5	1.8	28	1200
6	Apple	USA	2.8	200	25	0.05	0.6	2.2	40	700
7	BP	UK	1.5	300	15	0.07	0.4	1.3	20	1100
8	Total	France	2.3	180	18	0.05	0.3	1.6	30	900
9	Google	USA	3.0	190	20	0.06	0.7	2.1	35	750
10	Shell	Netherlands	2.1	250	12	0.08	0.2	1.4	15	950
11	Facebook	USA	2.9	180	22	0.05	0.8	2.3	38	800
12	EDF	France	2.4	160	30	0.04	0.4	1.7	25	1000
13	Amazon	USA	3.2	210	18	0.06	0.9	2.0	42	700
14	Microsoft	USA	2.7	170	24	0.05	0.5	2.5	37	750
15	IBM	USA	2.0	150	20	0.06	0.3	1.8	33	800
16	Coca-Cola	USA	2.1	160	10	0.07	0.2	1.2	22	900
17	McDonald's	USA	1.8	140	8	0.08	0.1	1.4	18	1000
18	Boeing	USA	1.6	130	5	0.09	0.2	1.3	12	950
19	Intel	USA	2.3	170	18	0.05	0.3	1.6	30	800

(continued)

Table 1 (continued)

No.	Company	Country	Value added by industry (%)	Carbon emissions (MtCO$_2$)	Renewable energy consumption (%)	Energy efficiency	Population growth rate (%)	Inflation rate (%)	Forest area (% of land)	Annual precipitation (mm)
20	Nissan	Japan	1.9	160	15	0.06	0.2	1.1	60	2100
21	Sony	Japan	2.2	180	17	0.05	0.4	1.0	70	2200
22	Honda	Japan	2.0	170	16	0.05	0.3	0.9	55	1900
23	Mitsubishi	Japan	1.7	150	14	0.06	0.2	1.0	65	2000
24	Panasonic	Japan	2.1	160	20	0.05	0.3	1.1	50	1800
25	Hitachi	Japan	2.3	170	22	0.04	0.4	1.2	45	1700
26	Toshiba	Japan	1.8	140	18	0.05	0.2	1.3	40	1600
27	Nestle	Switzerland	2.5	120	8	0.08	0.3	1.0	30	900
28	Novartis	Switzerland	2.2	110	5	0.09	0.1	1.5	25	800

Source Authors development using [29, 30]

```
    Example data
    clear
    input str10 company str20 country float Value_added_by_industry float
 Carbon_emissions float Renewable_energy float Energy_efficiency
    "Tesla" "USA" 3.5 250 30 0.05
    "Siemens" "Germany" 2.0 150 40 0.04
    "Toyota" "Japan" 1.8 180 20 0.06
    "Vestas" "Denmark" 2.5 120 60 0.03
    "Enel" "Italy" 2.2 140 35 0.04
    end

    * Calculate Value added by industry ( Value_added_by_industry _rate) and carbon
 emissions per capita
    gen Value_added_by_industry _rate = Value_added_by_industry
    gen Carbon_emissions_per_capita = Carbon_emissions

    * Display the results
    list company country Value_added_by_industry Carbon_emissions_per_capita
```

Fig. 2 Example data from Stata. *Source* Based on [28–30]

Industries, firms, and countries with higher RE and EE levels tend to have lower carbon emissions. For example, Siemens in Germany has lower carbon emissions than firms and nations with less formalization of sustainable energy. Japan, with a relatively large forest area (average 55%) and a low population increase rate (average 0.3%), has developed less carbon emissions than countries with small forest areas (average 32%) and a higher population increase rate (average 2).

In this consideration, the author focuses on illustrating how the promotion of renewable energy has benefitted economic enhancement and environmental conservation. The control variables, such as renewable energy consumption and energy efficiency, demonstrate a positive correlation with value added by industry and a negative correlation with carbon emissions among companies and countries. This implies that other sustainable energy practices are significant in providing economic development and overcoming environmental pollution. Government and the firms and organizations should therefore attempt to adopt and promote sustainable energy development policies to realize the economic development and environmental conservation.

The body of knowledge that stretches across the literature about the development and management of sustainable energy development strategies provides a valuable presentation of how policies and plans and initiatives toward sustainable energy could be designed and implemented. Through a focus on economic, social, and environmental frameworks and indicators, the model points out the way forward for ministers, managers, and citizens to transform the energy systems of the world.

5 Discussion

5.1 Framework for Sustainable Energy Development Strategy

An integrated approach to creating strategies for energy development and their further effective management directly affects the critical spheres of sustainable environment and economic and social advancements [32]. These include integrating renewable energy, implementing energy efficiency, appropriating the right policies, and planning for the right stakeholders' engagements which the authors consider to be the structural framework that need to be put in place [33].

Thus, the framework starts with a critical focus on integration of renewable energy, which bin underlining importance of the renewable energy sources including wind, water, and solar power. This comprises the establishment of goals on the uptake of renewable energy and policies or measures to encourage investments in renewable energy technologies.

Energy efficiency measures are also included as another feature of the framework because it is possible to markedly decrease the consumption of energy and emissions of CO_2 by improving the energy efficiency of societies [34]. This includes installing energy-efficient technology, encouraging energy conservation and carrying energy audits frequently.

Policy frameworks are central to catalyzing sustainable energy development [35]. The framework underlines the importance of policies for the change toward sustainable energy. This involved the creation of frameworks for the regulation of renewable sources of energy, offering incentives toward financing renewable energy technologies, and the creation of encouragement of research and development in sustainable energy technologies.

Another component of the identified framework is the engagement of stakeholders; the implication of sustainable energy solutions calls for collective effort between the government, business, and civil society. This comprises the participation of the concerned parties in the decision-making process, sensitization, and mobilization of partnerships regarding projects in electricity generation and utilization for sustainable energy.

Developing and managing sustainable energy development strategies is crucial with the help of eight fundamental elements presented in the framework derived from the literature review. Applying this framework will help appease policymakers, businesses, and other stakeholders in developing sustainable energy, hence improving the energy ecology for future generations.

One of the major components of the suggested framework for developing and managing sustainable energy development strategies is flexibility and adaptability, which are highly important when it comes to the changing nature of technologies and policies in the given field. Technological advancement requires a very dynamic system that can incorporate new units and new renewable energy system technologies every now and then. By including flexibility, policymakers and business entities can adapt to changes in the energy market, thus delivering the required flexibility to their strategies.

Besides, the very framework of conflict management should also be sensitive to changes in policy settings. Energy, environmental, and sustainability policies are dynamics in the contemporary globalized world due to scientific discoveries, political ideologies, and other societal factors. Thus, adapting changes can be easily included in an adaptable framework, making it possible for sustainable energy strategies to align with the current policy landscape and regulations.

Managing volatility hence boosts the robustness of sustainable energy plans as they may be affected by climate change effects and shifts in geopolitical powers. The approach ensures that flexibility is incorporated into the framework for application in the event of contingencies and obstacles that are liable to occur during a project's life cycle; the consideration of flexibility aims to achieve long-term sustainability of undertakings in producing sustainable energy.

This paper has demonstrated that flexibility and adaptability are inherent characteristics of the framework for developing sustainable energy development strategies.

5.2 Case Studies

Energy development as a part of sustainable development is essential for solving environmental challenges and enhancing economic growth [25]. Table 2 also captures a systemized framework of strategy formulation and management of sustainable energy development strategies concerning the 15 analyzed real-life cases for 2018–2023. These case studies, including the firms from different sectors, demonstrate the best-sustained energy practices and their effects on the financial and ecological results. Specifically, it is intended to serve as a conceptual map to help policymakers and businesses make better choices relevant to sustainable energy and getting well on their way to a better future.

The findings presented in fifteen case studies offer significant information regarding the effects of sustainable energy expansion policies on the ecological and economic results. Companies from different sectors and regions that target reducing energy consumption and promoting renewable energy sources mainly reveal positive outcomes.

For instance, Tesla has a high value added by industry of 3% because it produces renewable energy and energy efficiency. This emphasizes the fact that the use of sustainable energy activities has a direct link with economic development. Likewise, Siemens and Vestas companies also depict increased value added by industry, equal to 4.0% and 3.8%, respectively, as they emphasized renewable energy consumption and energy efficiency. Therefore, the results of this research imply that those companies participating and investing in sustainable energy practices can grow economically and minimize carbon emissions, thus assuming an environmentally sustainable position within business.

Table 2 Comprehensive overview of 15 real-world case studies

№.	Company	Data	Description	Clarifications
1	Tesla	Renewable energy consumption is 30% Energy efficiency is 0.05 Value added by industry is 3.5%	Tesla continues to promote the use of renewable energy and energy efficiency, hence lowering carbon emissions while the value added by industry constantly rises	The data is revealing the company's approach to the use of sustainable energy and the benefits it has on economic and solutions
2	Siemens	Renewable energy consumption is 40% Energy efficiency is 0.04 Carbon emissions is − 20%	In the case of the particular company, Siemens has been providing sufficient amounts of sustainable energy with markedly lower emission of carbon and significantly high efficiency level	The findings concern specifically Siemens' best practices regarding sustainable energy and the respective outcomes
3	Vestas	Renewable energy consumption is 60% Energy efficiency is 0.03 Value added by industry is 4.0%	Thus, Vestas' direction toward the use of renewable energy and energy efficiency has not only decreased carbon emissions' intensity but also increased the corporate value added to GDP	The information presented proves Vestas' effective practices in the field of the use of sustainable energy solutions
4	Toyota	Renewable energy consumption is 25% Energy efficiency is 0.06. Carbon emissions is − 15%	Sustainable energy improvements have been realized by Toyota in their energy efficiency reductions, and carbon emissions	The data also emphasizes Toyota's desire to make changes for the better and the continual improvement of environmental aspects
5	Google	Renewable energy consumption is 100% Energy efficiency is 0.02 Carbon emissions is − 50%	The commitment of Google for 100% renewable item has brought a drastic cut more than carbon foot printing	This information proves Google as the pioneer of sustainable energy measures
6	IKEA	Renewable energy consumption is 70% Energy efficiency is 0.04 Value added by industry is 3.8%	The lessons learnt from IKEA specifically the utilization of renewable energy and enhanced energy efficiency has had a positive bearing in executing the energy strategy toward cutting on carbon footprint as well as enhancing on economic development	This resulted in the following data showing IKEA's sustainable energy practices and the impact as illustrated below

(continued)

Table 2 (continued)

№.	Company	Data	Description	Clarifications
7	Apple	Renewable energy consumption is 75% Energy efficiency is 0.03 Carbon emissions is − 25%	Apple more than a decade in line with environmental protection has successfully reduced carbon emission and enhancing energy efficiency	The fact and figures discussed present Apple's sustainable energy activities and their implications
8	Unilever	Renewable energy consumption is 50% Energy efficiency is 0.05 Value added by industry is 3.7%	The policies formulated and implemented in the Unilever Company in relation to energy have greatly reduced carbon emission and enhanced economic advancement	Much of data shows the company's focus on sustainability and respect in its activities
9	Amazon	Renewable energy consumption is 40% Energy efficiency is 0.04 Carbon emissions is − 20%	Amazon has shown that through sustainable energy it has been able to reduce on its carbon footprint and enhance its energy use	The data at the tableau reveals Amazon's developments in sustainable energy initiatives
10	Microsoft	Renewable energy consumption is 80% Energy efficiency is 0.03 Value added by industry is 3.9%	Thus, the utilization of renewable energy and energy-efficient technologies by Microsoft has reduced the carbon emissions and increased the value added by industry	The above information reveals the sustainable energy status of Microsoft Corporation and its undertakings
11	Coca-Cola	Renewable energy consumption is 30% Energy efficiency is 0.06 Carbon emissions is − 15%	Coca-Cola Company energy management plans have led to the reduction of carbon emission and enhancement of energy utilization	The data helps to explain the picture that shows the company Coca-Cola's activities are aimed at environmentally friendly ones
12	BP	Renewable energy consumption is 20% Energy efficiency is 0.05 Value added by industry is 3.5%	The sustainable energy utilization program by the BP has achieved considerable success and helped in cutting down on the emission of carbon and subsequently have a positive effect on the economy	This information proves the specified BP's advancements in the management of sustainable energy

(continued)

Table 2 (continued)

№.	Company	Data	Description	Clarifications
13	Shell	Renewable energy consumption is 15% Energy efficiency is 0.04 Carbon emissions is − 10%	Shell addresses social issues like low carbon sentiments with regards to energy consumption and has pioneered on the use of renewable energy, hence cutting down its carbon footprints while at the same time promoting energy efficiency	The data focuses on sustainable energy measures and the company's activities in this area
14	Walmart	Renewable energy consumption is 50% Energy efficiency is 0.03 Value added by industry is 3.6%	Sustainable energy practices that Walmart has undertaken has indicated that they are useful since they have reduced emission of carbon and impacted the economic factors positively	This information also shows specifically how Walmart is working toward sustainability
15	Ford	Renewable energy consumption is 40% Energy efficiency is 0.05 Carbon emissions is − 20%	Sustainable energy initiatives at the Ford's company have in the recent past established a record of reducing carbon emission and energy efficiency	The information provided on Ford highlights the company's advancements in the utilization of green power resources

Source Authors development using [28–30]

Organizations such as Google and IKEA that set their goals to use only renewable energy, by 100% and 70%, respectively, exemplify considerable emissions decrease. For example, Google has cut its carbon emissions by half, clearly illustrating that setting and accomplishing high renewable energy targets is more than possible.

Specifically, the case study assessment of fifteen cases in the text highlights the relevance of sustainable energy activities for economic development as well as the protection of the environment. Companies committed to unsinging renewable energy and energy-efficient forums reap economic benefits and are also part of the world's effort to mitigate global warming.

5.3   Implementation and Management

Stakeholder management and policy coordination in achieving sustainable energy development involves engaging policy/process-based and robust monitoring measures to support the strategy [26]. All these components are essential for guaranteeing the success of practices that adopt sustainable energy, which drives economic progress and environmental conservation.

The key message derived from this is that stakeholder engagement is an important concept forming Sustainable Energy Development's basis [28]. Some measures that should be taken when it comes to stakeholder management should include consultation where stakeholders are involved in the initial planning of sustainable energy projects through public hearings, seminars, and workshops.

Freeman's fourth condition is also applicable, where stakeholder communication covering activity progress, issues, and achievements should be reported in different forms and forums, such as periodic and project reports and face-to-face and virtual meetings [18]. There is awareness raising and capacity development, which equips stakeholders with the proper knowledge and skills and comprises training programs, awareness-creating campaigns, and seminars. Further, partnerships with crucial stakeholders are developed as it improves the efficiency of sustainable energy projects due to more knowledge, resources, and credibility through a collaboration of governmental and non-governmental organizations, universities, and institutions, as well as industries involved in sustainability projects.

Policy coherence is crucial for establishing a favorable context for sustainable energy development [20]. Regulating and applying financial and non-financial reward power like tax credits, subsidies, grants, low-interest loans, fast-track permits, and award programs may promote investment in resourceful energy. The integration element regarding sustainable energy policies regarding economy and development policies such as industrial, agricultural, and urban development ensures a correct approach. The above approach is relatively more straightforward in achieving changes as there are numerous formations of intergovernmental relations in policy implementation involving agencies within the government through intertrack forces, working groups, and committees.

As necessary is the issue of monitoring mechanisms where sustainable energy development projects should be adequately monitored to ascertain their effectiveness [14]. The strategies include establishing performance goals and benchmarks to track aspects such as power production, improvement, emissions, and the economic consequences.

The audits and reviews performed by other employing third-party organizations with professional experience in sustainable energy ventures give an impartial look at its performance and conformity. Using adaptive management, one can readily accommodate such changes and respond proactively to changing circumstances by constantly adapting these strategies in light of new information, stakeholder feedback, and technology innovation.

This paper shows that through stakeholder engagement, policy coordination, and enhanced monitoring mechanisms efficient management of sustainable energy development strategies can be realized. Promoting a sustainable energy development plan involves a collaborative effort on the part of national, corporate, and local level society. All of them have their own specific and unique position as the key actors for the success and continuance of these activities.

Governments are central in developing and formulating policies and regulatory measures for energy growth for sustainability. The laws and regulations that can be made include legislation concerning the utilization of renewable energy forms, legislations that include strategies and measures for cutting down on carbon emissions, and the granting of incentives like tax credits, fellowships, and subsidies for investing in renewable energy technologies. Also, governments can sit down to fund corporate research and development to develop enhanced new technologies and structures for renewable energy sources. They are also involved in coordinating global cooperation and sourcing funds to digress and put policies on climate change and sustainable energy, which helps them set the standards to encourage other nations to emulate.

Applying sustainable energy is an essential goal for enterprises, and they are the main actors in the development of new solutions in this sphere [2]. They have the capital and know-how to design and implement futuristic mechanisms that more efficiently use energy than conventional fossil fuel-based ones. Applying the concept of sustainability in their operations, firms save costs and preserve their products' image. It is also probable to note that companies are also engaged in corporate social responsibility through offering financing for the production of non-renewable resources and collaborate with other enterprises, the authorities, or non-governmental organizations to increase the scope of such initiatives. Companies can also pressure supply chains to practice sustainable business, increasing the impact of business on the environment.

The roles of communities are crucial in the exploitation of renewable energy for their benefit [7]. There are local activities that countries can undertake, such as promoting a community-based renewable energy resource that, together with offering clean energy, also fuels the local economy by generating employment. Community sensitization and outreach about the possibilities of using sustainable energy can go a long way in changing people's attitudes and general mentality to embrace sustainable energy products in their homes and firms. Furthermore, the communities may be of utmost importance in offering feedback to policymakers and businesses to guarantee that the sustainable developments of energy projects address community needs and are introduced in socially sensible ways.

6 Conclusions

The major conclusions of this study were raised before people and highlighted the significance of the complex system strategy in the sphere of sustainable energy. An intensive examination of case studies of different companies and their location helps to determine that changing priorities toward renewable energy consumption and energy efficiency lead to substantial economic and environmental profits.

It is illustrated in the value added by industry data that Tesla, Siemens, Vestas, and other energy-consuming companies support the hypothesis that renewal energy consumption and energy efficiency cause the companies to have a reduced carbon footprint. For instance, Tesla's gross revealed was 3.5%. Besides, Siemens had 5% of the value added by industry, 50% from renewable energy sources, and the company scored 4. This means having a zero percent value added by industry and a sixty percent renewable energy ratio. Such conclusions leave room for sustainable consumption of energy practices to reinforce economic readiness in the face of climate change. According to estimations depicted in the model, up to a 10% rise in renewable energy consumption leads to a 1% rise. A target is to raise the contribution rate of value added by industry by 2% and reduce carbon emissions by 5%. For instance, Google has set a goal of receiving 100% renewable energy and acquired was 3%.

Thus, the corresponding results include a 2% value added by industry and a 50% reduction in carbon emissions, coinciding with the proposed model. Likewise, with a 70% renewals target, IKEA suffered a 2.8% decrease in the value added by industry and a 40% reduction in carbon emissions. These results show the quantitative outcomes of implementing sustainable energy and emphasize the necessity of effective strategies based on them. In terms of engaging stakeholders, considering policies and focusing on enhanced monitoring, sustainable energy development can meaningfully influence economic development and decrease the negative impact on the environment. Thus, to enhance the effects and reach in the overall area of sustainable energy, future research and policy making should consider the following.

In addition, the current literature should contain more information on integrating different renewable energy sources. Multiple types of research should be conducted to understand how integrating solar, wind, hydro, and biomass technologies enhances the provision of energy with storage security. This includes matters concerning energy storage systems such as batteries that help address the variability of renewable energy sources. Secondly, there should be research on the consequences of energy sustainability on individuals and societies' socioeconomic status.

This includes the impact on employment generation, economic development, social justice, and coping capabilities of a community. Awareness of such effects can assist in the optimization of the policy effects, primarily on vulnerable groups. Thirdly, it is necessary to introduce models for financing sustainable energy projects. Investigations should target factors such as green bonds and other PPPs to reduce the entry barriers and ensure the affordability of sustainable solutions. It is relevant for policies to focus on forming favorable legal regulations (policies to guide investors on how to operate, renewable energy targets, and financial tools such as tax credits).

Lastly, looking at such future trends, blockchain will help advance energy efficiency and its management to make energy consumption more sustainable. As the next step, further research and policies should consider the integration of renewable energy sources and the effects on socio-economy, innovative financial models, supportive regulations, cooperation on the international level, and the potential of new technologies. The result is at the junction of what is possible and can provide a more sustainable, equitable, and resilient energy future if the identified areas are managed.

References

1. M. Tymoshenko, V. Saienko, M. Serbov, M. Shashyna, O. Slavkova, The impact of industry 4.0 on modelling energy scenarios of the developing economies. Fin. Credit Activity Probl. Theor. Pract. **1**(48), 336–350 (2023). https://doi.org/10.55643/fcaptp.1.48.2023.3941
2. F.V. Bekun, A.A. Alola, S.A. Sarkodie, Toward a sustainable environment: nexus between CO_2 emissions, resource rent, renewable and nonrenewable energy in 16-EU countries. Sci. Total. Environ. **657**, 1023–1029 (2019)
3. M. Dudek, I. Bashynska, S. Filyppova, S. Yermak, D. Cichoń, Methodology for assessment of inclusive social responsibility of the energy industry enterprises. J. Clean. Prod. **394**, 136317 (2023)
4. K. Karpus, Not easy to 'green' old ways: national courts and rights-based smog cases in Poland. Rev. Euro. Compar. Int. Environ. Law **32**(1), 149–157 (2023)
5. M. Murshed, Are trade liberalization policies aligned with renewable energy transition in low and middle-income countries? An instrumental variable approach. Renew. Energy **151**, 1110–1123 (2020)
6. Z. Ahmed, M.M. Asghar, M.N. Malik, K. Nawaz, Moving towards a sustainable environment: the dynamic linkage between natural resources, human capital, urbanization, economic growth, and ecological footprint in China. Resour. Policy **67**, 101677 (2020)
7. O. Candra, A. Chammam, J.R.N. Alvarez, I. Muda, H.Ş. Aybar, The impact of renewable energy sources on the sustainable development of the economy and greenhouse gas emissions. Sustainability **15**(3), 2104 (2023). https://ssrn.com/abstract=4788972
8. N.P. Canh, N.T. Thong, Nexus between financialisation and natural resources rents: empirical evidence in a global sample. Resour. Policy **66**, 101590 (2020)
9. H. Jia, S. Fan, M. Xia, The Impact of renewable energy consumption on economic growth: evidence from countries along the belt and road. Sustainability **15**(11), 8644 (2023)
10. X. Liu, W. Zhang, X. Liu, H. Li, The impact assessment of FDI on industrial green competitiveness in China: based on the perspective of FDI heterogeneity. Environ. Impact Assess. Rev. **93**, 106720 (2022)
11. X. Ren, Y. Li, F. Wen, Z. Lu, The interrelationship between the carbon market and the green bonds market: evidence from wavelet quantile-on-quantile method. Technol. Forecast. Soc. Chang. **179**, 121611 (2022)
12. M. Shahbaz, M.A. Nasir, E. Hille, M.K. Mahalik, UK's net-zero carbon emissions target: investigating the potential role of economic growth, financial development, and R&D expenditures based on historical data (1870–2017). Technol. Forecast. Soc. Change **161**, 120255 (2020)
13. X. Xiao, L. Wang, Y. Wu, Y. Song, Z. Chen, X. He, Cathode regeneration and upcycling of spent LIBs: toward sustainability. Energy Environ. Sci. **16**(7), 2856–2868 (2023). https://doi.org/10.1039/dyyyyyyy
14. S. Zuo, M. Zhu, Z. Xu, J. Oláh, Z. Lakner, The dynamic impact of natural resource rents, financial development, and technological innovations on environmental quality: empirical evidence from BRI economies. Int. J. Environ. Res. Public Health **19**(1), 130 (2021). https://doi.org/10.3390/ijerph19010130
15. L.C. Muller, Hydrothermal liquefaction of spent coffee grounds followed by biocatalytic upgrading to produce biofuel: a circular economy approach. Biofuels (2021). https://doi.org/10.1080/17597269.2021.1948757
16. P. Bhola, Techno-economic and environmental assessment of utilizing campus building rooftops for solar PV power generation. Int. J. Green Energy (2021). https://doi.org/10.1080/15435075.2021.1904946
17. O. Prokopenko, A. Chechel, A. Koldovskiy, M. Kldiashvili, Innovative models of green entrepreneurship: social impact on sustainable development of local economies. Econ. Ecol. Socium **8**, 89–111 (2024). https://doi.org/10.61954/2616-7107/2024.8.1-8
18. O. Prokopenko, T. Kurbatova, M. Khalilova, A. Zerkal, G. Prause, J. Binda, T. Berdiyorov, Y. Klapkiv, S. Sanetra-Półgrabi, I. Komarnitskyi, Impact of investments and R&D costs in

renewable energy technologies on companies' profitability indicators: assessment and forecast. Energies **16**(3), 1021 (2023). https://doi.org/10.3390/en16031021

19. G. Trypolska, T. Kurbatova, O. Prokopenko, H. Howaniec, Y. Klapkiv, Wind and solar power plant end-of-life equipment: prospects for management in Ukraine. Energies **15**(5), 1662 (2022). https://doi.org/10.3390/en15051662

20. R. Li, Y. Xin, I. Sotnyk, O. Kubatko, I. Almashaqbeh, S. Fedyna, O. Prokopenko, Energy poverty and energy efficiency in emerging economies. Int. J. Environ. Pollution **69**(1–2), 1–21 (2022). https://doi.org/10.1504/IJEP.2021.125188

21. L.Y. Dovhan, H.A. Mokhonyko, I.P. Malyk, *Project Management: A Textbook for Studying the Discipline for Masters of the Field of Knowledge 07 "Management and Administration" Specialty 073 "Management" Specialization: "Management and Business Administration", "Management of International Projects", "Management of Innovations", "Logistics"* (KPI Named After Igor Sikorsky, Kyiv, 2017)

22. T. Kurbatova, D. Lysenko, G. Trypolska, O. Prokopenko, M. Järvis, T. Skibina, Solar energy for green university: estimation of economic, environmental and image benefits. Int. J. Glob. Environ. Issues **21**(2–4), 198–216 (2022). https://doi.org/10.1504/IJGENVI.2022.126209

23. I. Sopronenkov, N. Zelisko, V. Vasylyna, I. Lutsenko, V. Saienko, Tax policy: impact on business development and economic dynamics of the country. Econ. Aff. **68**(04), 2025–2034 (2023). https://doi.org/10.46852/0424-2513.4.2023.14

24. V. Koval, Y. Sribna, S. Kaczmarzewski, A. Shapovalova, V. Stupnytskyi, Regulatory policy of renewable energy sources in the European national economies. Polityka Energetyczna Energy Policy J. **24**(3), 61–78 (2021). https://doi.org/10.33223/epj/141990

25. V. Koval, N. Savina, Y. Sribna, L. Filipishyna, D. Zherlitsyn, T. Saiapina, European energy partnership on sustainable energy potential. IOP Conf. Ser. Earth Environ. Sci. **1126**(1), 012026 (2023). https://doi.org/10.1088/1755-1315/1126/1/012026

26. L. Todorov, A. Aleksandrova, T. Ismailov, Relation between financial literacy and carbon footprint: review on implications for sustainable development. Econ. Ecol. Socium **7**(2), 24–40 (2023). https://doi.org/10.31520/2616-7107/2023.7.2-2

27. Y. Sribna, S. Skakovska, T. Paniuk, I. Hrytsiuk, The economics of technology transfer in the environmental safety of enterprises for the energy transition. Econ. Ecol. Socium **7**(1), 84–96 (2023). https://doi.org/10.31520/2616-7107/2023.7.1-8

28. World Bank, *World Bank Open Data* (2024). https://data.worldbank.org/. Accessed 15 May 2024

29. IMF, *IMF Finance Data* (2024). https://www.imf.org/en/Data. Accessed 15 May 2024

30. IEA, *Energy Statistics Data Browser* (2024). https://www.iea.org/data-and-statistics. Accessed 15 May 2024

31. V. Mazur, A. Koldovskyi, L. Ryabushka, N. Yakubovska, The formation of a rational model of management of the construction company's capital structure. Fin. Credit Activity Probl. Theor. Pract. **6**(53), 128–144 (2023). https://doi.org/10.55643/fcaptp.6.53.2023.4223

32. A. Kuczabski, O. Aleinikova, H. Poberezhets, H. Tolchieva, V. Saienko, A. Skomorovskyi, The analysis of the effectiveness of regional development management. Int. J. Qual. Res. **17**(3), 695–706 (2023). https://doi.org/10.24874/IJQR17.03-05

33. M. Masyk, Z. Buryk, O. Radchenko, V. Saienko, Y. Dziurakh, Criteria for governance' institutional effectiveness and quality in the context of sustainable development tasks. Int. J. Qual. Res. **17**(2), 501–514 (2023). https://doi.org/10.24874/IJQR17.02-13

34. I. Bashynska, L. Niekrasova, V. Osypov, A. Dyskina, L. Zakharchenko, Conceptual basis for the formation of a smart eco-industrial parks as benchmarking of sustainable manufacturing. In *The 17th International Conference Interdisciplinarity in Engineering. Inter-ENG 2023. Lecture Notes in Networks and Systems*, 928, 337–349, eds. by L. Moldovan, A. Gligor (Springer, Cham, 2024)

35. N. Shmygol, O. Galtsova, K. Shaposhnykov, S. Bazarbayeva, Environmental management policy: an assessment of ecological and energy indicators and effective regional management (on the example of Ukraine). Polityka Energetyczna **24**(4), 43–60 (2021)

Evaluating Waste Heat Recovery Potentials and Key Performance Indicators in Energy-Intensive Sectors

Larissa Wurster and **Viktor Koval**

Abstract Energy efficiency and climate protection have shaped environmental discussion. Climate change objectives and regulations are increasing, especially in heat- and energy-intensive industries. An example is the iron and steel industry, which accounts for approximately 4–5% of global energy consumption and CO_2 emissions. This study reviews waste heat recovery potentials and elucidates methodologies for their quantification based on physical and technical optima while ensuring economic production conditions (cost minimization, maximization of plant utilization, preservation of product quality, and competitiveness). It further introduces and explains the KPIs as specific heat input (kWh/t) and heat utilization rate (%), describing the economic potential of waste heat, overall prices, and the cost structure of alternative production technologies or heat sources. These KPIs are critical for evaluating waste heat utilization efforts and detailing their calculation processes. Ways of waste heat recovery are linked to an overview of possible technologies.

Keywords Energy efficiency · Waste heat recovery · Energy-intensive sectors · Process heat · Economic potential

L. Wurster (✉)
Daimler Truck AG, 10 Fasanenweg, 70771 Leinfelden-Echterdingen, Germany
e-mail: wurster.larissa@gmail.com

V. Koval
Izmail State University of Humanities, 12 Repina Str., Izmail 68610, Ukraine

1   Introduction

1.1   Legal Background of Climate Policy and Efficiency Challenges

Energy efficiency and climate protection play significant roles in modern society. Keywords such as climate neutrality and decarbonization are increasingly becoming the focus of corporate policy discussion. In addition to the need to tackle climate change and the responsibility toward future generations, further tightening climate targets and EU legislation play an essential role. One example is the European Green Deal, which requires greenhouse gas savings of 55% by 2030 compared with 1990 [1]. Securing the energy supply is a vital issue that became apparent at the beginning of the war in Ukraine, and the fear of energy bottlenecks due to the gas supply embargo of sanctioned countries [2–4].

Further examples of the growing importance of environmental legislation and regulation are the Corporate Sustainability Reporting Directive (CSRD) [5] and the Energy Performance of Buildings Directive [6]. The German counterpart, the so-called "Gebäudeenergiegesetz (GEG)," also focuses on waste heat recovery [7]. Energy is needed in one of its possible supply sources in all sectors. Figure 1 shows the final energy consumption of EU-27 in 2019 based on the shares of the fuels used.

Approximately 37% of consumption is based on oil and petroleum products, 23% on electricity, 21% on natural gas, and 11% on renewables and biofuels. The heat consumption of EU-27 was only 5%. Nevertheless, it is a significant consumer, especially in the industrial sector. This is underlined by the world heat generation, as shown in Fig. 2.

The world's heat generation makes it clear that most of the required heat is not generated by renewable energy sources (5.7%). Most of the heat was generated by gas

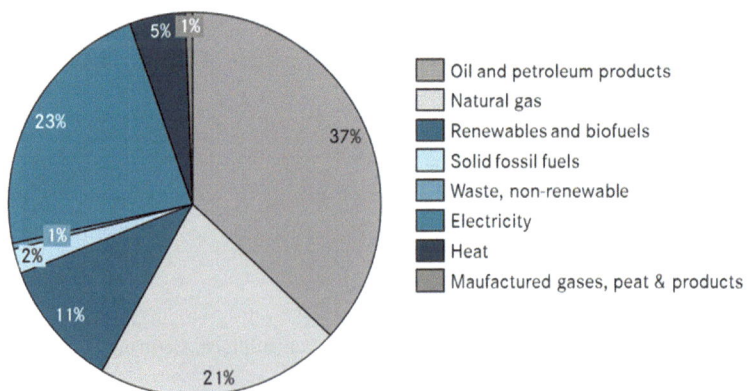

Fig. 1   Final energy consumption by fuel for the EU-27 (2019). *Source* Based on [8]

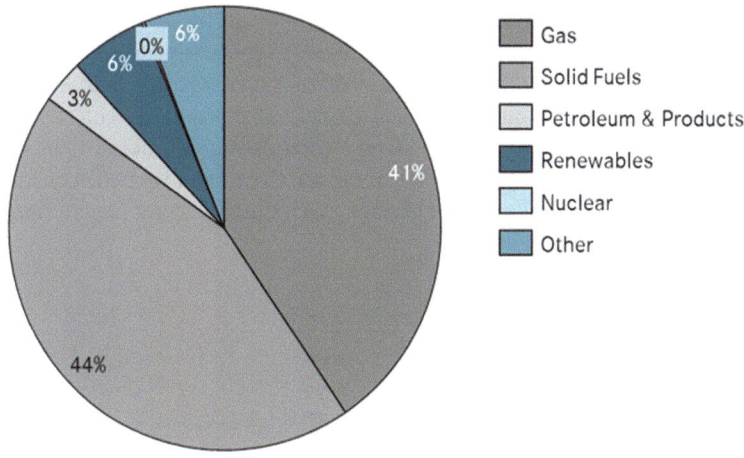

Fig. 2 World heat generation by fuel. *Source* Based on [8]

(40.6%) or solid fuels (44.3%). Figure 3 shows the energy consumption of different sectors, such as industry, household, and transport in the European Union (EU-27).

With 31%, transport is the most significant player in energy consumption in the EU, with 26% being households and industry in the second place. The numbers in the United States are comparable; the industrial sector "accounts for approximately one-third of all energy used in the United States" [10].

New legal requirements are also challenging for industrial companies while maintaining competitiveness and reducing energy costs. The challenges associated with climate change mitigation, such as reducing energy costs and the availability of renewable energy sources, are difficult to overcome in the energy- and heat-intensive

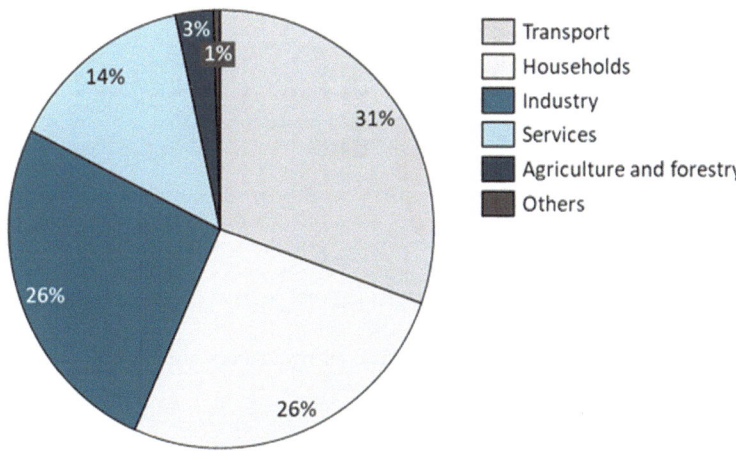

Fig. 3 Energy consumption by sectors in the EU-27 (2018/2019). *Source* Based on [8, 9]

industries. This also underlines the final energy demand for industrial applications and the different energy sources used to produce the process heat demand (Fig. 4).

Process heat plays a significant role in 66% of the energy demand. Most of it is produced with gas (45%), a quarter is made of coal, 9% is produced with electricity or district heating, and only 5% is made with renewable energy. Studies [16, 17] mention a similar heat demand (73%) for the UK industry. Compared with the consumption process in the European Union (Fig. 1), heat in the industrial sector plays a significant role.

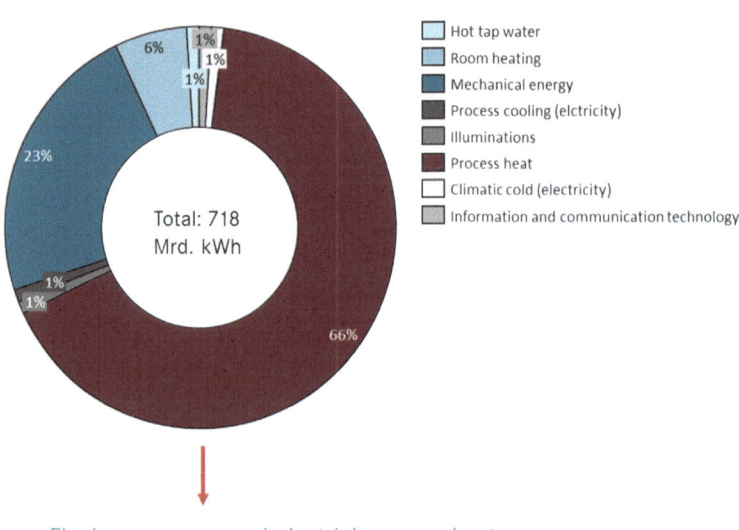

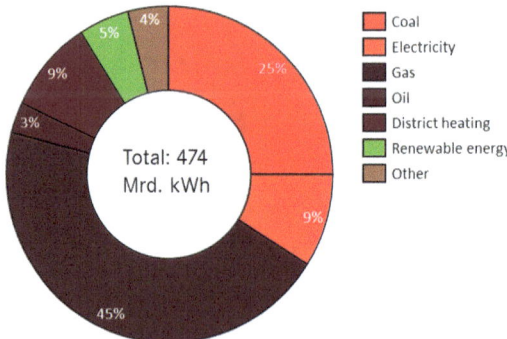

Fig. 4 Final energy demand in industry and final energy sources of industrial process heat in Germany. *Source* Based on [11–15]

1.2 Energy- and Heat-Intensive Sectors

One industrial sector that requires a large amount of process heat is the energy- and heat-intensive. The steel and iron industry account for 4–5% of the world's energy demand and CO_2 emissions [18, 19]. There are around 560 foundries in Germany that offer approximately 75,000 jobs.

These account for approximately 30% of the CO_2 emissions of the German industry and approximately 60% of them for iron foundries [20, 21]. The preservation of the prospects of these production facilities is prevented by the prescribed CO_2 saving targets (e.g., the substitution of energy sources such as natural gas and coal) while ensuring economic production conditions (cost minimization, maximization of plant utilization, preservation of product quality, and competitiveness) [20, 22, 23]. Figure 5 summarizes the energy consumption of the German foundry industry.

Based on the research, 50% of the energy consumption of a foundry is generated with electricity, 33% with natural gas, and 17% with foundry coke.

Heat, in particular, and, above all, process warmth, is an essential lever in the foundry. This also shows the energy requirements of a light alloy foundry's production process chains (Fig. 6).

Around 60% of the energy consumption is attributable to melting and warm holding, and 55% of the electrical energy is used for process heat generation [20, 24].

Process heat demand mainly arises in melting operations, holding and heat treatment, and the pan-industry [22, 26]. "With a share of around 70 percent of the total energy costs, it represents a central starting point for measures to increase energy efficiency" [22]. But how could the energy efficiency be increased? One methodology to overcome these challenges is waste heat recovery. The study aims

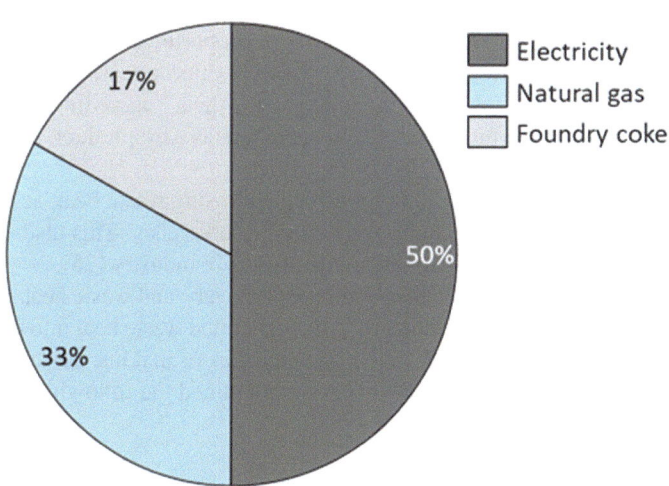

Fig. 5 Energy consumption of the German foundry industry. *Source* Based on [21, 24]

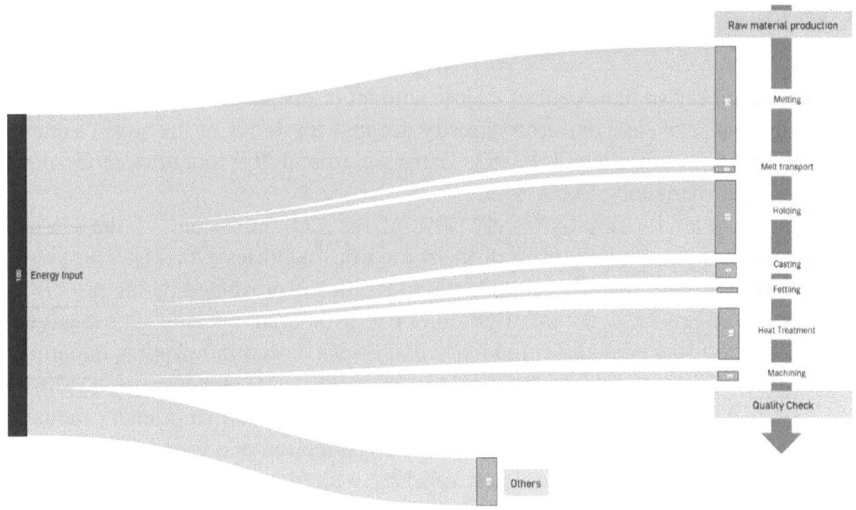

Fig. 6 Energy requirements in a light alloy foundry. *Source* Based on [25, 26]

to enrich the discourse on energy conservation and sustainability in industrial operations, contributing to the broader goal of reducing environmental impact through improved efficiency.

2 Literature Review

In energy- and heat-intensive industries, a large proportion of the resources used is waste heat after the actual process step, which is often released unused into the environment. This also underlines the term "waste heat" according to [27]. This defines waste heat as "warmth generated by a process as a by-product. It is often left unused but may still be useful" [27].

Temperature is crucial, especially for effectively using waste heat, as mentioned in [27]. The higher the temperature, the better the efficiency. This also underlines the typical possibilities of making waste heat usable in industry [28].

[29 in 13] divides the possible uses into heat recovery and waste heat utilization. Heat recovery refers to directly integrating the generated waste heat into production. Waste heat utilization focuses on various possibilities of making waste heat usable beyond the actual production process. This study defined the following three types:

1. For external use.
2. Internal use only.
3. Conversion to other energy sources (e.g., electricity and cooling).

The third aspect of waste heat conversion has been extensively researched [including 14, 30]. A wide range of waste heat conversion technologies are available in the market (see also Sect. 3.2). The conversion into electricity by the Organic Rankine Cycle (ORC) and the conversion into cooling by the absorption chiller should be highlighted. The integration of such technology is viral in energy- and heat-intensive industries (e.g., foundry). Here, waste heat usually has a high temperature level, albeit with significant differences in the individual process steps [14, 31].

However, the conversion of waste heat into other forms of energy is often not very lucrative, as a relatively low yield can be expected owing to the generally poor efficiency of individual technologies (e.g., ORC < 10%) [32].

Using waste heat in the form of warmth is significantly more lucrative. Research focuses on external use, i.e., disclosure to third parties [33, 34]. For example, concepts that have already been implemented usually include heating neighboring residential areas or nearby municipal properties, such as swimming pools. In some cases, waste heat is passed on to neighboring industrial companies. A good example is the waste heat utilization of a foundry in Singen, Germany. Here, the waste heat from a cupola furnace in thermal oil is delivered to the neighboring Maggie plant (Nestlé) and used to generate saturated steam [33].

The integration of waste heat into existing district heating networks (often under the responsibility of a local energy supplier) represents an opportunity for reliable waste heat utilization. The functioning of heat networks is, among other things, discussed in [35, 36]. It must be considered that district heating networks often require an even heat supply at a specified temperature level. The flexibilization of heat networks is the subject of [37, 38] and temperature-variable heat network deals [39].

In the case of significant waste heat sources, the local district heating network can be influenced by the waste heat source, such as unstable waste heat quantity and different temperature levels. This can result in the deterioration of the grid quality in the form of grid jumps or temperature fluctuations. The control of these undesirable, retroactive influences on the overall system must be ensured. A plant heat network is to be understood as a complex energy source that enormously influences a more extensive supply network, such as an energy sink. Therefore, optimizing a complex networked topology based on energy sources and sinks is challenging.

In [40], it deals with the regulated and unregulated integration of waste heat in decentralized energy supply structures and sets up a simulation model for this purpose. Further research is focused on integrating combined heat and power (CHP) plants into district heating networks. A stable heat supply is generally guaranteed here [38, 41].

Since a company that generates a high amount of waste heat usually has a very high heat requirement, the focus should be on the internal use of the waste heat [28]. The problem here is often the integration of warmth. This is optional and close to the actual place of origin. If available, transport to other buildings/areas can be done via the heating system. If this is designed similarly to a district heating network as a plant heating network, distribution of the waste heat is relatively easy to implement, as significantly lower heat losses are generated than with other forms of transport [42].

3 Waste Heat Recovery Potential-Method

3.1 Principles and Applications of Industrial Waste Heat Utilization

Important to know is, that nearly every production process generates heat or waste heat [15]. How much depends on the process itself and the efficiency of it.

For the use of waste heat, [13, 15] set out the basic principles, which are shown in Fig. 7.

The best way is not to generate waste heat—avoiding, reducing, or minimizing heat generation. Because something that does not get generated has not to be converted or reused. Especially [15] recommends:

1. Avoidance of obvious excess consumptions (e.g., sensible shutdown, avoidance of safety surcharges).
2. Reduction of applicable energy requirements (e.g., thermal insulation, process and procedure optimization).
3. Improvement of energy utilization (e.g., optimize utilization, use coupling effects).

Another option is energy recovery, for instance, in the process itself. Therefore, [13] differs between the process returns either to the same (2.1) or lower (2.2) temperature level. Here, the efficiency ratio is much higher than the third option—waste heat utilization conversion.

The basic principles for waste heat utilization [13, 15] are expanded by [29 in 13] to include the usage options of the actual waste heat use (Fig. 8).

Waste heat recovery is also circular because waste heat can be used more than once. An indirect usage at a lower temperature level is also possible [15].

There are two options for using waste heat in the industrial sector: heat recovery systems and waste heat utilization [13, 29]. As mentioned, it generates nearly every production process (waste) heat. The best and most efficient way to use this waste

Fig. 7 Basic principles for waste heat utilization. *Source* Based on [13, 15]

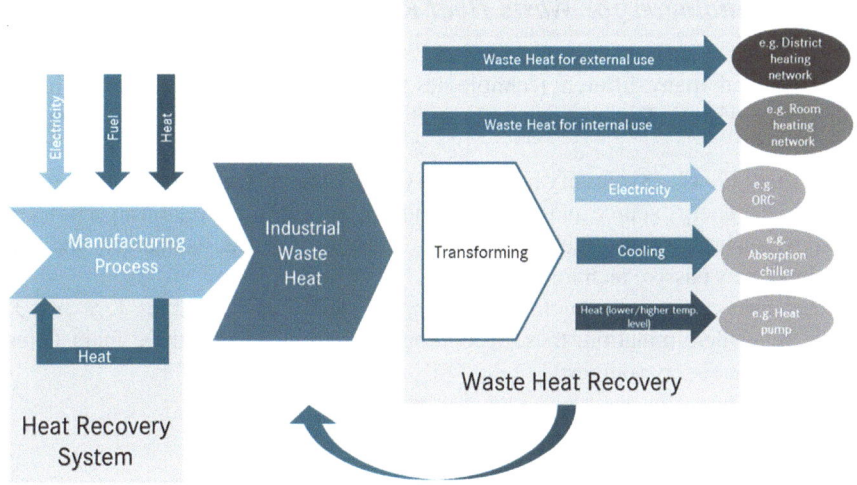

Fig. 8 Potential uses of industrial waste heat. *Source* Based on [13, 15, 29]

heat is directly inside the manufacturing process. This is defined as a heat recovery system [13, 29]. An excellent example could be the preheating of input materials or media such as combustion air [43].

All heat from this first option is defined by [13, 29] as industrial waste heat. This industrial waste heat can either be unused or recovered (waste heat recovery). For the waste heat recovery, it also mentions three different options:

- Waste heat for external use.
- Waste heat for internal use.
- Transformation to other energy sources.

Waste heat for external or internal use can be at the same or a higher/lower temperature level. Typical examples are the integration in a (local) district heating network or a room heating network on the production side.

The third option is the transformation to other energy sources. Common are electricity or cooling. Efficiency is strongly connected to temperature level and availability.

Besides the definition of [13] mentioned [44], waste heat can be used in various ways. Three types of waste heat utilization can be distinguished: direct heat utilization, conversion into other energy sources, and heat storage [44]. Direct utilization and conversion are also mentioned in [29]. The point of heat storage was added, but it is also state of the art.

Many technologies are available for heat storage and transformation, which will be described shortly.

3.2 Technologies for Waste Heat Recovery

As mentioned, many different technologies for waste heat recovery are available on the market [45, 46]. They can generally be classified as passive or active technologies [17, 47, 48].

Studies [17, 47, 48] classify heat recovery technologies, which transform the heat into another energy source and bring it to another temperature level during a process, as active technology and technologies like heat storage, heat exchanger, and heat distribution as passive technologies.

The most common technologies are shortly described in (Tables 1, 2 and 3). For a better determination, these technologies are divided into three main fields, better-saying energy sources:

• Transformation from heat to heat (Table 1);
• Transformation from heat to cooling (Table 2);
• Transformation from heat to electricity (Table 3).

Tables 1, 2 and 3 only show an abstract of the most common technologies and not a complete market overview. In Tables 2 and 3, the active technologies [17, 47, 48] are considered exclusively.

Table 1 Technologies for the "transformation" from heat to heat

Technology	Power range	Heating/ Cooling media	Temperature level (waste heat)	COP
Heat pump, e.g., compression heat pump	15–20,000 kW	Air, liquid	< 50 °C	3–5
Heat exchanger, e.g., rotary heat exchanger	< 1600 kW	Gas	< 650 °C	–
Heat storage, e.g., latent heat storage	< 2.5 MWh	Liquid, salt solutions	< 0–400 °C	–

Note COP Coefficient of performance
Source Based on [15, 49]

Table 2 Technologies for the transformation from heat to cooling

Technology	Power range	Heating/Cooling media	Temperature level	Heat ratio
Absorption chiller	10–6000 kW	Cooling media and solvent	70–130 °C	0.6–0.7

Source Based on [15, 22, 50]

Table 3 Technologies for the transformation from heat to electricity

Technology	Power range	Heating/Cooling media	Temperature level	Efficiency ratio
Organic Rankine cycle (ORC)	10–10,000 kW	Organic working fluid	70–450 °C	10–20%
Stirling engine	< 50 kW	Working medium	100–700 °C	13–36%

Source Based on [13, 47, 48, 51–53]

3.3 Potentials of Waste Heat Recovery

A lot of literature on the potential of waste heat recovery has been published over the years.

For example, [10] described ways to improve industrial energy efficiency in the US and mentioned that 20–50% of the consumed energy in manufacturing processes is getting lost but that waste heat recovery systems can improve energy efficiency by 10–50% [10].

A study [54] focuses on industrial waste heat in Germany. "88% of the waste heat (111 PJ/a) is produced within only six sectors." [54] According to [54], these sectors are the metal industry, glass and ceramic, and the chemical sector. The estimate by [54] is a theoretical potential since no aspects of technical usability are included in the estimation [51].

Study [55] mentions a theoretical potential of 223 PJ (62 TWh) in exhaust gas flows for Germany, which is calculated based on data from offering memorandums. Another example is [56], which calculates a waste heat potential for Germany of 12% (depending on the final industrial energy consumption above 140 °C) by deriving parameters based on a Norwegian study from [57].

The work of [30] shows that the main waste heat potential can be found in the steel and iron sector (2.7 PJ/a technical potential). A study [58] mentions that around 40% of the total energy consumption of a foundry could be recovered via waste heat recovery if suitable heat sinks are available.

One of the main challenges in calculating theoretical waste heat potentials is the need for more data, which results in estimations that depend on the final energy consumption [30]. To overcome this challenge and to ensure compatibility, key figures are required. An example of this can be the physical optimum, according to [59].

A study [15] mentions that much information is required to calculate potentials. For example, the medium of the waste heat, including density ρ (kg/m^3) and specific heat capacity c_p (kJ/(kg K)), as well as the temperature T (°C or K) and the volume flow $\dot{V}$ (m^3/s).

This is supplemented by [18], which mentions other important criteria for waste heat utilization: operating hours (for yearly potentials), temperature level (for the usage options), carrier medium (influences type and efficiency), and time course.

Depending to available temperature levels, that can be classified into three categories [48]:

- High temperature level: > 650 °C.
- Medium temperature level: 232–649 °C.
- Low temperature level: < 232 °C.

3.4 Methodologies for Quantification

The basic principles of the methodologies, according to [59], are used to compare the different potentials and calculate one's potentials. This methodology is also used for the Key Figure Generation (Sect. 3.4.3). The calculations from [60, 61] are also used in this article, focusing on energy and heat-intensive sectors, especially the foundry areas. Some basic questions about the methodologies are, first of all:

- What is meant by the technical optimum or the physical optimum?
- How can different systems be compared with each other?
- Where is the physical optimum of a process, and where is it technically feasible?
- What literature is available to create the basis for determining potential?

3.4.1 Physical Optimum

According to [59], the physical optimum (PhO) refers to the theoretically optimal reference process, the optimal state. Physical, biological, and chemical conditions are considered and used for the ideal thermodynamic comparison process [59]. Figure 9 shows this relationship again.

One of the advantages of the physical optimum is that this reference point is regarded as ultimate and cannot be undercut by further developments [59].

According to [59], there are no "losses due to non-stationary driving style (partial load losses): unplanned downtime, partial load losses of the system, interruption for cleaning and maintenance; product scrap (rework, animal feed, waste); heat losses on the system surface (convection and heat radiation); dissipation losses (friction); changes in potential and kinetic energy related to product flow; unwanted chemical changes to the product; mass defects".

The physical optimum corresponds to the maximum available potential that could be achieved under idealized conditions.

Fig. 9 Physical optimum (PhO). *Source* Based on [59]

3.4.2 Technical Optimum

This is contrasted with the potential, referred to below as the technical optimum (TO), which is assessed based on the latest state of the art technology currently available. According to [60], "the technical optimum serves as a reference value to determine the feasible technical improvement potential" and "corresponds to the currently best available technology (BAT)" [60].

Figure 10 is intended to illustrate the different efficiency levels of physical and technical optimum. Here, the physical optimum is also differentiated from the Perpetuum Mobile. The Perpetuum Mobile is a process that cannot be implemented in reality, as it refutes the principles of thermodynamics [59].

Figure 10 vividly illustrates the formidable challenge of achieving the physical optimum in reality, a state that is similar to the Perpetuum Mobile. The ideal conditions necessary for this optimum are not only demanding but also impossible to be fully and simultaneously realized.

The state of the art, a goal that holds immense potential, could be a more feasible aspiration in the production environment. This is where the latest scientific findings and a research environment can truly revolutionize production processes. The best available technologies, when harnessed within a production, can lead to significant advancements. It is crucial for a company to be advised and to proactively develop its production to incorporate these technologies. The state of the art corresponds to the technical optimum and is a good reference value or benchmark for a production environment.

A distinction must be made between technically usable, theoretically usable, and economically usable potential (Fig. 11).

Figure 11 also adds the economic potential to take economic factors like costs, taxes, and laws into account. For that, an economic analysis needs to be conducted. Furthermore, it understood the physical optimum as the theoretical potential and the technical optimum as the technical potential. Technical limitations and demand analyses limit the technical potential analysis [67, 68].

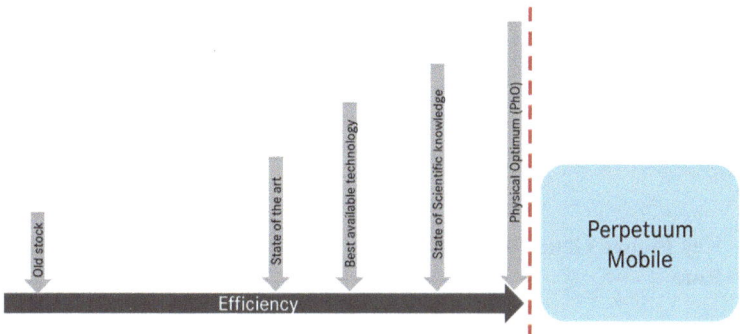

Fig. 10 Efficiency levels of physical and technical optimum compared with the Perpetuum Mobile and current state of the art. *Source* Based on [59]

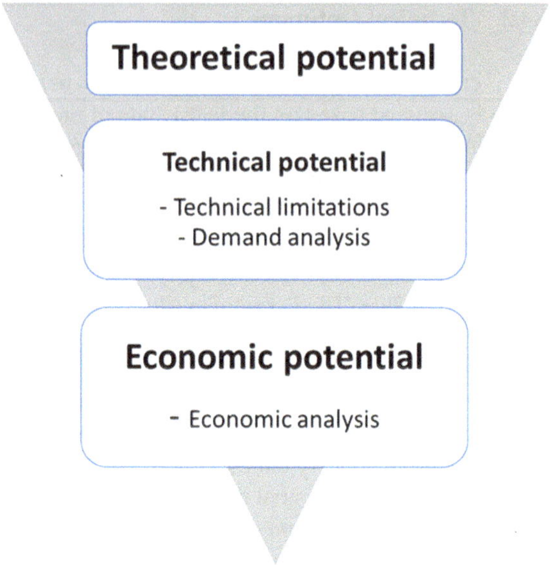

Fig. 11 Theoretical, technical, and economic potential compared. *Source* Based on [54, 55, 62–66]

A process that generates waste heat means that the theoretically usable waste heat potential is understood to be the "theoretically decouplable waste heat quantities under the assumption of a constant, minimum heat sink temperature (neglecting technical restrictions)" [55].

The technically usable waste heat potential is the "(external, pipe-bound) dischargeable waste heat quantities considering 'real' heat sinks and temperatures as well as the corresponding technologies to be used" [55].

The economically usable waste heat potential refers to the "proportional technical waste heat potential, which takes general prices and cost structures of alternative generation technologies into account or heat sources that can be developed and used economically" [55].

The physical optimum, which belongs to [59] and the thoughts about it of [60, 61], focuses on the foundry sector compared with the technical optimum orientated on the work of [54]. Important to know is, that the physical optimum is the ideal reference process. In contrast, actual production orientates more on benchmarks, like the foundry of Fondium in Singen (Germany) [33], or the technical optimum, closely connected to the state-of-the-art and the best available technologies (BAT) [69].

3.4.3 Key Figure Generation—Specific Heat Input and Heat Utilization Rate

To compare different (waste) heat potentials (especially in sectors like the foundry industry), the key performance indicator (KPI)-*specific heat input in* kWh per ton (here liquid iron) orientated on [60, 61] gets introduced.

[61] uses the KPI of *specific energy consumption* [kWh/t] which is formed from the "ratio of the supplied electrical energy [kWh] to the produced quantity of casting material [t]" [61].

However, in the present work, the amount of recovered heat (heat recovery) [kWh] is set in relation to the melted tonnage—here iron [tFe] and is referred to as specific heat input [kWh/tFe]. In order to determine the key figure, both the total *heat recovery quantity* [MWh or kWh] and the *melted tonnage* [tFe] must be known.

The following calculation formula is used:

$$\text{specific heat input} \left[\frac{\text{kWh}}{\text{t}_{\text{Fe}}} \right] = \frac{\text{heat recovery [kWh]}}{\text{melted tonnage [t}_{\text{Fe}}]}$$

This key figure is used to determine the individual potentials and compare them with each other.

Another KPI is the heat utilization rate [%], which compares the amount of used heat content of a process with the unused. For that, the energy input and the energy sources' heat content or calorific value are needed, e.g., the heat content of coke in a cupola furnace, to which is referred in the following. A similar KPI energy recovered [%] is also presented in [70].

In the calculation, the energy input, here for example of the coke, gets compared with the recovered energy, here the used waste heat. The amount recovered and the melted tonnage of the aggregate, for example, a cupola furnace, should be assumed for the calculation of the heat utilization rate.

1. Melted tonnage [t_Fe].

The melted tonnage is required for the calculation and can, for example, be taken from sustainability reports or from other data sources which belongs to the investigated aggregate. It is also possible to calculate the molten tonnage if the input of used energy sources, for example, electricity, and the level of utilization of this source are known.

2. Heat demand of used energy source here as an example the coke demand [kWh/t_Fe].

For the determination of heat demand, it is assumed that one ton of liquid iron consists of 10.5% coke [71]. This corresponds to 105 kg (0.105 t) of coke per t_Fe. This must now be set in relation to the heat content of the coke of 29 MJ/kg or 8.06 kWh/kg [72].

The combination of the heat content and the amount of coke lead to the heat content per ton of liquid iron.

$$\text{heat content per t\_Fe} \frac{[\text{kWh}]}{[\text{t}_{\text{Fe}}]} = \text{heat content of coke} \left[\frac{\text{kWh}}{\text{kg}} \right]$$
$$* \text{amount of coke} \left[\text{kg/t} \right]$$

Resulting in:

$$8.06 \, \frac{kWh}{kg\_coke} * 105 \, \frac{kg_{coke}}{t_{Fe}} = 846.3 \, \frac{kWh}{t\_Fe}$$

This means, that to produce 1 t_Fe, a coke input of 846.3 kWh is required.

3. Max. heat content [MWh]

The melted tonnage [t_Fe] used in the first step is now linked to the heat demand of the used energy source of the second step, to determine the maximum heat content (optimum depending on the melting capacity). If known, here also benchmarks or the physical or technical optimum can be used instead of the heat demand of the used energy source.

$$\frac{\text{melted tonnage } [t_{Fe}] * \text{heat content per } t_{Fe} \left[\frac{kWh}{t_{Fe}} \right]}{1000} = \text{max. heat content } [MWh]$$

4. Used vs. unused heat content [%]

For the used heat content, the amount of heat recovery/heat recovery quantity [MWh] is needed. The maximum heat content [MWh] is set in relation to the heat recovery quantity [MWh] to determine the used heat content [%].

$$\frac{\text{heat recovery quantity } [MWh]}{\text{max. heat content } [MWh]} * 100\% = \text{used heat content } [\%]$$

The used heat content [%] is offset by the unused heat content [%].

$$1 - \frac{\text{heat recovery quantity } [MWh]}{\text{max. heat content } [MWh]} * 100\% = \text{unused heat content } [\%]$$

5. Loss of heat [MWh]

In addition to the used and unused heat content, the amount of heat loss can be determined.

$$\text{max heat content } [MWh] - \text{heat recovery quantity } [MWh] = \text{loss of heat } [MWh]$$

3.4.4 Example Foundry

To gain a better understanding, the two KPIs *specific heat input* and *heat utilization rate* are shown with some example figures. If an output of 250,000 t_Fe from an example foundry gets used and 60,000 MWh/a amount of energy provided by waste heat are assumed, the following calculation will lead to the result:

$$\text{specific heat input} \left[\frac{\text{kWh}}{\text{t}_{\text{Fe}}} \right] = \frac{60,000 \text{ MWh/a}}{250,000 \text{ t\_Fe}} * 1.000$$

$$\text{specific heat input} \left[\frac{\text{kWh}}{\text{t}_{\text{Fe}}} \right] = 240 \frac{\text{kWh}}{\text{t}_{\text{Fe}}} / \text{a}$$

To show the calculations belonging to the heat utilization rate, the shown example starts at the third step. For the first step, the molten iron from the calculation of the specific heat input [kWh/t_Fe] of 250,000 t_Fe is used and the coke rate of 10.5%, which leads to a heat content per t_Fe of 846.3 kWh/t_Fe, as well as the heat recovery quantity of 60,000 MWh; out of that, the max. heat content [MWh] gets calculated.

$$\frac{250,000 \, [\text{t\_Fe}] * 846.3 \, [\text{kWh/t\_Fe}]}{1000} = 211,575 \, [\text{MWh}]$$

Next, the used and unused heat content [%] get calculated, as well as the loss of heat [MWh].

Used heat content [%]:

$$\frac{60,000 \, [\text{MWh}]}{211,575 \, [\text{MWh}]} * 100\% = 28.36 \, [\%]$$

Unused heat content [%]:

$$1 - \frac{60,000 \, [\text{MWh}]}{211,575 \, [\text{MWh}]} * 100\% = 71.64 \, [\%]$$

Loss of heat [MWh]:

$$211,575 \, [\text{MWh}] - 60,000 \, [\text{MWh}] = 151,575 \, [\text{MWh}]$$

The heat utilization rate is also shown in Fig. 12.

The amount of used or unused heat content strongly depends on the chosen baseline. Here, the maximum heat content. The factors change if a benchmark or the physical optimum gets used as a baseline. If the heat utilization rate of more than one year or aggregate gets compared, the available optimization potential or gaps can be shown. Since the total amount of heat recovered in a foundry depends heavily on the amount of molten iron, which varies significantly over time and depending on the furnace size, it is challenging to compare different furnaces and years. Therefore, the shown KPIs can be used to ensure that one unique calculation method is used.

Fig. 12 Visualization of the heat utilization rate

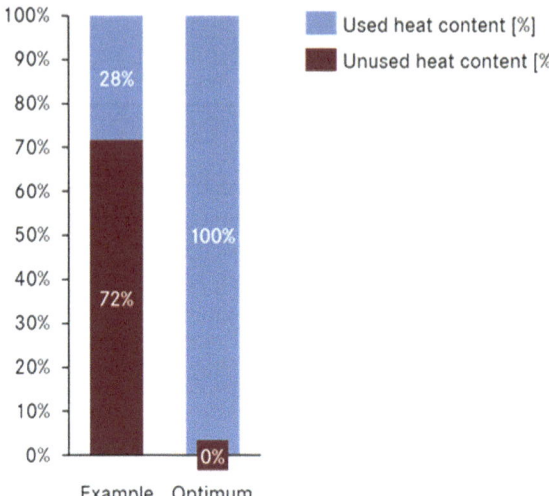

4 Conclusions

The study shows relevant laws and legislation regarding environmental and energy efficiency, such as the European Green Deal, which requires substantial decreases in energy consumption and CO_2 emissions. The EU's energy consumption is also shown, and it is broken down into different industrial sectors. It then focuses on the industrial process of heat demand and explains the effects of regulations on the energy- and heat-intensive industries.

As an example of energy- and heat-intensive industries, this study focuses more on the foundry sector and shows the waste heat potential. Relevant literature regarding different potentials and calculation methods are presented. Furthermore, waste heat recovery methods and technologies for applying applicable waste heat are explained. Various technologies are classified depending on their temperature levels and energy sources. The article also breaks down the physical and technical optimum into two KPIs relevant to energy- and heat-intensive sectors. It presents the specific heat input [kWh/t_Fe] and introduces the heat utilization rate [%]. Both KPIs were shown within a reference process to gain a better feeling. Based on this, a specific heat input of 240 kWh/t_Fe and a heat utilization rate of 28% resulted in a melted tonnage of 250,00 t_Fe and a heat recovery quantity of 60,000 MWh.

Further work should focus on integrating other heat- and energy-intensive industries and consider different waste heat recovery technologies. The possible usage and direct transfer to other energy sources closely connected to heat networks should be investigated in more detail.

References

1. European Commission (ed.), *The European Green Deal: striving to be the first climate-neutral continent*. https://commission.europa.eu/strategy-and-policy/priorities-2019-2024/european-green-deal_en
2. Bundesnetzagentur (ed.), *Versorgungssicherheit* (2022). https://www.bundesnetzagentur.de/DE/Fachthemen/ElektrizitaetundGas/Versorgungssicherheit/start.html
3. R. Guccione, Italian foundries risk: production stoppage because of the war in Ukraine. A&L (2), 66–67 (2022)
4. T. van de Sand, Von der Krise in die Krise. Giesserei. Die Zeitschrift Für Technik, Innovation und Management **109**(5), 80–83 (2022)
5. Directive (EU) 2022/2464 of the European Parliament and of the Council of 14 December 2022 amending Regulation (EU) No 537/2014, Directive 2004/109/EC, Directive 2006/43/EC and Directive 2013/34/EU, as regards corporate sustainability reporting (2022). https://eur-lex.europa.eu/legal-content/EN/TXT/?uri=CELEX:32022L2464
6. Directive (EU) 2024/1275 of the European Parliament and of the Council of 24 April 2024 on the energy performance of buildings (2024). https://eur-lex.europa.eu/eli/dir/2024/1275/oj
7. Gesetz zur Einsparung von Energie und zur Nutzung erneuerbarer Energien zur Wärme- und Kälteerzeugung in Gebäuden (Gebäudeenergiegesetz - GEG) (2023). https://www.gesetze-im-internet.de/geg/BJNR172810020.html
8. European Commission (ed.), *EU Energy in Figures: Statistical Pocketbook 2021* (Publications Office of the European Union, 2021)
9. Energy, transport and environment statistics: 2020 edition, 2020 edn. *Statistical Books*, vol. 2020 (Publications Office of the European Union, 2020)
10. I. Johnson, T. William, W.T. Choate, A. Amber Davidson, *Waste Heat Recovery: Technology and Opportunities in U.S. Industry*, ed. by U.S. Department of Energy (2008)
11. Agentur für erneuerbare Energien (ed.), *Endenergieträger für industrielle Prozesswärme* (2020). https://www.unendlich-viel-energie.de/mediathek/grafiken/endenergietraeger-fuer-industrielle-prozesswaerme
12. A. Aydemir, C. Rohde, *What About Heat Integration? Quantifying Energy Saving Potentials for Germany* (Industrial Efficiency, Berlin, 2018, June 13)
13. J. Hesselbach, *Energie- und klimaeffiziente Produktion: Grundlagen, Leitlinien und Praxisbeispiele* (1. Aufl.) (Springer Vieweg, Berlin, 2012)
14. S. Hirzel, B. Sonntag, C. Rohde, *Industrielle Abwärmenutzung: Kurzstudie* (2013)
15. J. Strack, *Technologien der Abwärmenutzung* (2016). https://www.saena.de/download/broschueren/BU_Technologien_der_Abwaermenutzung.pdf
16. G.P. Hammond, J.B. Norman, Heat recovery opportunities in UK industry. Appl. Energy **116**, 387–397 (2014). https://doi.org/10.1016/j.apenergy.2013.11.008
17. J.I. Chowdhury, Y. Hu, I. Haltas, N. Balta-Ozkan, G. Matthew Jr., L. Varga, Reducing industrial energy demand in the UK: a review of energy efficiency technologies and energy saving potential in selected sectors. Renew. Sustain. Energy Rev. **94**, 1153–1178 (2018). https://doi.org/10.1016/j.rser.2018.06.040
18. M. Junge, Demonstration eines strukturierten, simulationsgestützten Vorgehens zur Konzeption der Abwärmenutzung in der Stahlindustrie am Beispiel der Georgsmarienhütte GmbH. Abschlussbericht des Forschungsprojekts (2011)
19. H. Zhang, H. Wang, X. Zhu, Y.-J. Qiu, K. Li, R. Chen, Q. Liao, A review of waste heat recovery technologies towards molten slag in steel industry. Appl. Energy **112**, 956–966 (2013). https://doi.org/10.1016/j.apenergy.2013.02.019
20. Deutsche Energie-Agentur GmbH (ed.), *Leuchttürme CO2-Einsparung Industrie: Gießerei-Industrie* (2022). https://www.co2-leuchttuerme-industrie.de/branchen/branchensteckbrief-giesserei-industrie/
21. Haus der Gießerei-Industrie (ed.), *Energiebedarf der deutschen Gießerei-Industrie*. https://www.guss.de/prozess/aktuelles-aus-der-verbandsarbeit/innoguss/energiebedarf-der-deutschen-giesserei-industrie

22. Deutsche Energie-Agentur GmbH (ed.), Systematische Energieeffizienz steigern und CO2-Emissionen senken in der Gießerei-Industrie [special issue] (2021)
23. S. Otto, I. Heinrich, Einsatz eines geeigneten Energiedatenmanagementsystems zur Effizienzsteigerung energieintensiver Industrieunternehmen, pp. 91–107
24. M. Bosse, DIN ISO 50003: Neue Pflicht zum Nachweis von Energieeffizienzmaßnahmen in Gießereien: Aktuelle energiepolitische Rahmenbedingungen und Auswirkungen der neuen 50000er-Normenfamilie. Giesserei **105**(2), 89–91 (2018)
25. S. Scharf, B. Sander, M. Kujath, H. Richter, E. Riedel, H. Stein, J. Tom Felde, FOUNDRY 4.0: an innovative technology for sustainable and flexible process design in foundries. Procedia CIRP **98**, 73–78 (2021). https://doi.org/10.1016/j.procir.2021.01.008
26. S. Scharf, M. Kujath, B. Sander, H. Seidel, E. Riedel, W. König, C. Michealis, J. Volkert, J. tom Felde, N. Stein, H. Stein, Innovatives Technologie- und Anlagenkonzept für eine nachhaltige Prozessgestaltung in Gießereien. Giesserei -Spezial (1), 78–85 (2019)
27. R. Paschotta, *Abwärme* (2010, Mar 7). https://www.energie-lexikon.info/abwaerme.html
28. M. Pehnt (ed.), *Energieeffizienz: Ein Lehr- und Handbuch* (1., korr. Nachdr) (Springer, Berlin, 2010). https://doi.org/10.1007/978-3-642-14251-2
29. F. Leischner, Vorgehensweise, Potenziale und Möglichkeiten für die Nutzung industrieller Abwärme. Diplomarbeit, Universität Kassel, Kassel (2011)
30. L. Grote, P. Hoffmann, G. Tänzer, Abwärmenutzung - Potentiale, Hemmnisse und Umsetzungsvorschläge. Studie (2015)
31. M. Blesl, K. Hufendiek, P. Radgen, *Abwärmepotentiale in der Industrie: Konzepte zur Nutzung im Mittel- und Niedrigtemperaturbereich* (1. Auflage) (Beuth Verlag, 2022)
32. R. Paschotta, *Organic Rankine Cycle* (2014, Dec 19). https://www.energie-lexikon.info/organic_rankine_cycle.html
33. F. Bruch, *Abwärmenutzung eines Kupolofens: Vorhabennummer 20119 UBA-FB AP 20119.* Umweltinnovationsprogramm des Bundesministeriums für Umwelt, Naturschutz und Reaktorsicherheit (2009, June). http://www.uba.de/uba-info-medien/4133.html
34. C. Franzke, Überbetriebliche Reststoffverwertung und Abwärmenutzung in der Industrie. Dissertation, Technische Universität Berlin, Berlin. DataCite (2019)
35. MVV Netze - Ein Unternehmen der MVV, TAB-Heizwasser: Technische Anschlussbedingungen für Nah- und Fernwärme der MVV Netze GmbH (2015)
36. D. Quaggiotto, J. Vivian, A. Zarrella, Management of a district heating network using model predictive control with and without thermal storage. Optim. Eng. **22**(3), 1897–1919 (2021). https://doi.org/10.1007/s11081-021-09644-w
37. R. Ramos, F. Rojas, J. Becerril, M. Rebolledo, A. Gonzalez, metabolon—measuring flow and temperature on the heating network: Kompendium der Forschungsgemeinschaft (2015, May), pp. 65–78
38. F.C. Ertem, M. Acheampong, Impacts of demand-driven energy production concept on the heat utilization efficiency at biogas plants: heat waste and flexible heat production. Process Integr Optim Sustain **2**(1), 1–16 (2018). https://doi.org/10.1007/s41660-017-0024-z
39. M. Werner, M. Ehrenwirth, S. Muschik, C. Trinkl, T. Schrag, Operational experiences with a temperature-variable district heating network for a rural community. Chem. Eng. Technol. **151**, 889 (2021). https://doi.org/10.1002/ceat.202100114
40. T. Bornemann, Industrial Waste Heat Utilization: Spannungsfeld zwischen Abwärmenutzung und Kraft-Wärme-Kopplung in der produzierenden Automobilindustrie. Dissertation, Kassel University Press GmbH. DataCite (2018)
41. M. Stritzinger, Betriebssimulation der Integration eines BHKW in ein neu strukturiertes Wärmeversorgungssystem. Bachelorarbeit, Hochschule Karlsruhe - Technik und Wirtschaft, Karlsruhe (2011)
42. M. Sandrock, C. Maaß, S. Weisleder, H. Westholm, W. Schulz, G. Löschan, C. Baisch, H. Kreuter, D. Reyer, D. Mangold, M. Riegger, C. Köhler, *Kommunaler Klimaschutz durch Verbesserung der Effizienz in der Fernwärmeversorgung mittels Nutzung von Niedertemperaturwärmequellen am Beispiel tiefengeothermischer Ressourcen: Abschlussbericht.* CLIMATE CHANGE 31/2020. Umweltforschungsplan des Bundesministeriums für Umwelt, Naturschutz und nukleare Sicherheit (2018, Apr)

43. M. Sprecher, H.B. Lüngen, B. Stranzinger, H. Rosemann, W. Adler, *Abwärmenutzungspotenziale in Anlagen integrierter Hüttenwerke der Stahlindustrie: Abschlussbericht.* 07/2019. Umweltforschungsplan des Bundesministeriums für Umwelt, Naturschutz und nukleare Sicherheit (2018, Nov)
44. M. Feil, Abwärmenutzung am Nfz-Motor: Konstruktiver Gesamtentwurf – Motorprüfstand. Bachelorarbeit, Hochschule Ulm, Ulm (2011)
45. K. Biel, C.H. Glock, Prerequisites of efficient decentralized waste heat recovery and energy storage in production planning. J. Bus. Econ. **87**(1), 41–72 (2017). https://doi.org/10.1007/s11573-016-0804-x
46. C. Ononogbo, E.C. Nwosu, N.R. Nwakuba, G.N. Nwaji, O.C. Nwufo, O.C. Chukwuezie, M.M. Chukwu, E.E. Anyanwu, Opportunities of waste heat recovery from various sources: review of technologies and implementation. Heliyon **9**(e13590), 1–27 (2023). https://doi.org/10.1016/j.heliyon.2023.e13590
47. A. Anastasovski, P. Rasković, Z. Guzović, A review of heat integration approaches for organic Rankine cycle with waste heat in production processes. Energy Convers. Manage. **221**(113175), 1–17 (2020). https://doi.org/10.1016/j.enconman.2020.113175
48. S.O. Oyedepo, B.A. Fakeye, Waste heat recovery technologies: pathway to sustainable energy development. J. Therm. Eng. **7**(1), 324–348 (2021)
49. J. Barco-Burgos, J.C. Bruno, U. Eicker, A.L. Saldaña-Robles, V. Alcántar-Camarena, Review on the integration of high-temperature heat pumps in district heating and cooling networks. Energy **239**, 1–16, Article 122378 (2022). https://doi.org/10.1016/j.energy.2021.122378
50. EnergieAgentur.NRW, Kälteerzeugung: Potenziale zur Energieeinsparung (2010)
51. A. Aydemir, H. Doderer, F. Hoppe, S. Braungardt, *Abwärmenutzung im Unternehmen.* Studie für das Ministerium für Umwelt, Klima und Energiewirtschaft Baden-Württemberg (2019)
52. S. Broberg Viklund, M.T. Johansson, Technologies for utilization of industrial excess heat: potentials for energy recovery and CO2 emission reduction. Energy Convers. Manage. **77**, 369–379 (2014). https://doi.org/10.1016/j.enconman.2013.09.052
53. C. Persson, J. Olsson, Comparison of various co-generation technologies (Jämförelse mellan olika kraftvärmeteknologier): Report no. SGC 128 (2002)
54. S. Brückner, Industrielle Abwärme in Deutschland: Bestimmung von gesichertem Aufkommen und technischer bzw. wirtschaftlicher Nutzbarkeit. Dissertation, Technische Universität München, München (2016)
55. S. Blömer, Potenziale und Hemmnisse außerbetrieblicher Abwärmenutzung in Deutschland: Ergebnisse des Verbundvorhabens "EnEff: Wärme: Netzgebundene Nutzung industrieller Abwärme (NENIA)". BMUB. 4. BMUB-Fachtagung Klimaschutz durch Abwärmenutzung, Berlin (2018, Oct 18)
56. M. Pehnt, J. Bödeker, M. Arens, E. Jochem, Die Nutzung industrieller Abwärme – technisch-wirtschaftliche Potenziale und energiepolitische Umsetzung: Bericht im Rahmen des Vorhabens "Wissenschaftliche Begleitforschung zu übergreifenden technischen, ökologischen, ökonomischen und strategischen Aspekten des nationalen Teils der Klimaschutzinitiative". FKZ 03KSW016A und B (2010)
57. G. Sollesnes, H.E. Helgerud, Utnyttelse av spillvarme fra norsk industri: en potensialstudieenova (2009)
58. M. Putz, M. Cherkaskyy, A. Esche, C. Fanghänel, A. Schlegel, *Energieeffizienzpotenzial in der Planung am Beispiel der Gießerei-Industrie.* Abschlussbericht. Fraunhofer-Institut für Werkzeugmaschinen und Umformtechnik IWU (2015)
59. D. Volta, Das physikalische Optimum als Basis von Systematiken zur Steigerung der Energie- und Stoffeffizienz von Produktionsprozessen. Dissertation, Technische Universität Clausthal, Clausthal. DataCite (2014)
60. P. Wenzel, P. Radgen, J. Westermeyer, Energieeffizienter Induktionsofenbetrieb durch Annäherung an das physikalische Optimum. Giesserei **107**(7–8), 40–46 (2020)
61. P.M. Wenzel, Exergetische Analyse und physikalisches Optimum zur Bewertung und Steigerung der Ressourceneffizienz am Beispiel eines Mittelfrequenzofens. Masterarbeit, Universität Stuttgart, Stuttgart (2019)

62. M. Blesl, S. Kempe, M. Ohl, U. Fahl, A. König, T. Jenssen, L. Eltrop, Wärmeatlas Baden-Württemberg: Erstellung eines Leitfadens und Umsetzung für Modellregionen. Forschungs-bericht, Universität Stuttgart, Stuttgart (2008)
63. S. Brueckner, R. Arbter, M. Pehnt, E. Laevemann, Industrial waste heat potential in Germany—a bottom-up analysis. Energ. Effi. **10**(2), 513–525 (2017). https://doi.org/10.1007/s12053-016-9463-6
64. S. Brueckner, L. Miró, L.F. Cabeza, M. Pehnt, E. Laevemann, Methods to estimate the industrial waste heat potential of regions—a categorization and literature review: a categorization and literature review. Renew. Sustain. Energy Rev. **38**(38), 164–171 (2014). https://doi.org/10.1016/j.rser.2014.04.078
65. B. Metz, O. Davidson, P. Bosch, R. Dave, L. Meyer, Mitigation of climate change: the physical science basis contribution of Working Group I to the Fourth Assessment Report of the Inter-governmental Panel on Climate Change. Climate Change 2007: Working Group 3 (Cambridge University Press, 2007)
66. H. Roth, K. Lucas, W. Solfrian, F. Rebstock, Die Nutzung industrieller Abwaerme zur Fern-waermeversorgung: Analyse der Hemmnisse fuer die Nutzung industrieller Abwaerme zur Fernwaermeversorgung. Forschungsbericht 10407312 (No. 40) (1996)
67. H. Wolff, Innovative Techniken: Beste verfügbare Techniken (BVT) in ausgewählten indus-triellen Bereichen Teilvorhaben 3: Gießereien: Volume 3: Technikerhebung 2012. TEXTE 82/2014 (2013)
68. N. Savina, Y. Sribna, N. Pitel, L. Parkhomenko, A. Osipova, V. Koval, Energy management decarbonization policy and its implications for national economies. IOP Conf. Ser. Earth Environ. Sci. **915**(1), 012007 (2021). https://doi.org/10.1088/1755-1315/915/1/012007
69. Y. Kazancoglu, Y. Berberoglu, C. Lafci, O. Generalov, D. Solohub, V. Koval, Environmental sustainability implications and economic prosperity of integrated renewable solutions in urban development. Energies **16**(24), 8120 (2023). https://doi.org/10.3390/en16248120
70. J. Selvaraj, K.I. Ramachandran, D. Venkatesh, S. Devanathan, Greening the foundry sector by an innovative method of energy conservation and emission reduction, in *IEEE 8th International Conference on Intelligent Systems and Control (ISCO)* (2014), pp. 60–63. https://doi.org/10.1109/ISCO.2014.7103919
71. J. Würz, J. Rachner, Willich, H.-J. Rachner, M. Lemperle, Lohnt sich Abwärmenutzung beim Kupolofen? Giesserei **100**(03), 76–80 (2013)
72. AGRAR PLUS GmbH, *Heizwerte-/äquivalente* (2023). https://agrarplus.at/heizwerte-aequiv alente.html

Infrastructural and Marketing Support of the Energy Saving System: Strategic Guidelines for National Programs Recovery

Natalia Kuzmynchukⓘ, **Tetiana Kutsenko**ⓘ, **Hanna Pysarevska**ⓘ, **and Tetyana Hrebenyk**ⓘ

Abstract Increasing the energy efficiency level of buildings is crucial for recovering the Ukrainian economy and resolving the dependence on external energy sources. This study presents a toolkit of scientific and methodical approaches to the infrastructural and marketing support of energy-saving systems, allowing one to solve the existing problems in the national energy sector systematically. Based on monitoring the results, practical strategies for further development have been identified, aimed at constantly increasing energy efficiency, energy sustainability, economic security, and scaling up energy efficiency programs. It has been established that the technical and economic substantiation of energy-saving projects based on assessing energy costs and economic effect levels allows the creation of a foundation for involving new participants in scaling up positive experiences. The energy-saving system infrastructure and marketing support are vital in implementing national economy recovery projects by stimulating the demand for energy-efficient solutions and the transition to sustainable development. The practical implementation of this approach in the example of one communication and logistics infrastructure facility made it possible to reduce the cost of heat energy consumption by 5.8 times. The discounted payback period for the implemented energy-saving measures ranges from 0.16 to 3.16 years. Scaling this project will expand the influence of energy efficiency processes on other critical infrastructure facilities, increase the overall energy efficiency level in various economic sectors, reduce dependence on energy sources, and contribute to preserving natural resources and reducing their negative impact on the environment.

Keywords Energy efficiency · Energy saving · Infrastructural and marketing support

N. Kuzmynchuk · T. Kutsenko (✉) · H. Pysarevska
V. N. Karazin Kharkiv National University, 4 Svobody Sq., Kharkiv 61022, Ukraine
e-mail: chkutsenko@gmail.com

T. Hrebenyk
Separate Structural Department, Classical Professional College of Sumy State University, 39 Sadova Str., Konotop 41615, Ukraine

V. Koval (ed.), *Renewables in the Circular Economy and Business*,
SpringerBriefs in Applied Sciences and Technology,
https://doi.org/10.1007/978-3-031-72174-8_4

1 Introduction

Energy saving as a strategic priority in restoring the national economy of Ukraine involves identifying modern technologies and innovations that contribute to reducing energy consumption and increasing energy efficiency as the basis of the energy independence of business, community, and nation as a whole. As a result, it is crucial to examine the possible financial, technical, and social barriers and difficulties that could impede the successful rollout of energy-saving initiatives in Ukraine.

It was noted in [1] that energy efficiency stimulates the development of innovative technologies for the rational use of energy resources. For industrial buildings, energy efficiency means lower utility costs and increased productivity and asset value. Active energy systems have been suggested as a way to enhance the energy efficiency of industrial buildings. The utilization of artificial neural networks in creating a model to forecast energy usage, as suggested in [2], enables the exploration of different thermal properties. The model improves the quality of energy consumption planning and management and can also be applied to office buildings. Having an accurate estimate of energy consumption, which is validated by the strong correlation between predicted and actual usage, is crucial for making well-informed decisions on how to best optimize energy resources. Researchers [3] have noted the growing interest in finding the most effective methods and approaches to increase the energy efficiency of buildings. Here, the main problems and motivational factors affecting the implementation of energy-efficient solutions are identified.

The study [4] examined the issues related to the slow pace of implementing innovative energy technologies (SETs) in buildings despite their proven effectiveness in reducing energy consumption and increasing energy efficiency. The focus was on examining the reasons hindering these technologies' application. According to the survey results, the factors affecting the application of intelligent energy technologies were determined, and practical recommendations for speeding up their implementation were provided.

Strategic guidelines for Ukraine's national recovery programs include specific goals for reducing energy consumption, increasing energy efficiency in various sectors of the economy, and implementing programs to encourage energy savings through financial and tax incentives.

Moreover, mechanisms for monitoring and evaluating the achievement of these goals have been determined to ensure the effectiveness and transparency of the national energy conservation policy. Thus, according to the National Energy Efficiency Action Plan until 2030 [5], the volumes of primary and final energy consumption in Ukraine are expected to reduce by 22.3% and 17.1%, respectively. In turn, the Energy Strategy of Ukraine until 2050 [6] emphasizes the need to create conditions for the post-war restoration of the sustainable growth of the national economy by ensuring access to reliable, sustainable, and modern energy sources. By 2050, the energy sector must achieve a high level of climate neutrality, which involves access to clean energy, overcoming energy poverty, creating a modern decentralized energy system, ensuring the efficient functioning of domestic energy markets, and their integration into global markets.

The central values that guide Ukraine's energy strategy are economic viability, environmental sustainability, economic affordability, fair access to energy, and market compatibility. Hence, Ukraine possesses a considerable opportunity for energy efficiency, considering the importance of updating current infrastructure, building new energy-efficient structures, and adopting energy-efficient technologies in key sectors like construction, industry, transport, and services, thus opening up new avenues for business growth. The development of comprehensive approaches is necessary to support national programs on energy efficiency and energy saving. These approaches should take into account technical aspects, as well as social and economic factors, promote active participation, create educational programs, and establish financial support mechanisms to encourage investment in energy-saving initiatives. Therefore, it is important to focus on developing efficient strategies and practices for implementing energy-saving programs in Ukraine, considering global trends and international best practices, in order to contribute to the country's restoration and development.

The study aims to develop theoretical foundations and practical recommendations for developing the energy-saving system infrastructure and marketing security in the context of following strategic guidelines of national programs for restoring the economy of Ukraine and its regions.

2   Materials and Methods

A significant part of Ukraine's energy sector was seriously damaged: more than half of the thermal generation, 30% of solar generation, and 90% of wind generation had been either out of order or inaccessible. Moreover, several public mines have been closed. Many power lines, substations, and gas lines were damaged, negatively affecting households' access to electricity and gas. However, despite all the difficulties, a large-scale initiative for repairing the energy facilities was implemented in 2023. In less than a year, a capacity of 16 GW was ready for the heating season, including an extra 2.2 GW of thermal and hydro generation added to the energy system. The successful integration of the national energy system into the European system allowed Ukraine and Poland to increase their electricity exchange capacity by creating a joint 400-km-long electricity transmission line, thus increasing their import capacity to 1.7 GW. The national energy market is integrated into the European market by altering and implementing the updated provisions of the Regulation on Wholesale Energy Market Integrity and Transparency (REMIT) [7]. Draft laws have been intensively developed for inclusion and adoption in the Fourth Energy Package named "Clean Energy for All Europeans," together with the National Energy and Climate Plan. Certification of "Ukrtransgaz" JSC by the new rules established by the EU as the gas storage facilities operator will enable foreign traders to pump over 31 billion m^3 of gas from non-residents into storage facilities, which testifies to their trust and willingness to cooperate when gas storages in Europe operate at maximum capacity. Having completely deprived the oligarchs of control over gas distribution systems, 27 gas distribution companies became state-owned to ensure a non-stop gas supply and accept the challenges of today. The process of reorganizing "Energoatom" National Nuclear Energy Generating Company into a legal entity has already been completed, and the Supervisory Board for the Operator of Gas Transportation System of Ukraine has been created as a part of the corporate governance reform. These national energy companies play a crucial role as they are essential for implementing corporate reforms in line with global management standards. This will open up more opportunities for attracting investments and fostering growth in different economic sectors.

In addition, during the winter of 2023–2024, Ukraine only used its own domestically extracted gas for energy, leading experts to believe that the country is progressing toward achieving energy independence by 2024 and could potentially resume exporting electricity. For the first time in history, the volume of national gas extracted from the new drills exceeded 1 billion cubic meters per year, which was possible because of the launch of 86 new drills. This allows Ukraine to be transformed into a European clean energy hub and implement the tasks set in the Energy Strategy until 2050 [6].

Nevertheless, the main measure of energy efficiency, the energy intensity of Ukraine's GDP, is notably higher than that of advanced European countries due to inadequate technological advancements, an imperfect industrial structure in the national economy, and the influence of the shadow economy [8]. This scenario objectively limits the competitiveness of agricultural production and places a significant burden on the economy. Unlike other Western countries, where energy saving is considered a practical, economical, and environmentally friendly choice, Ukraine struggles to survive in conditions of vital necessity to restore the economy on the principles of providing energy efficiency and resolving the issue of dependence on external energy sources. Energy efficiency is a critical element of Ukraine's energy strategy. The primary goals of increasing energy efficiency and taking all the opportunities for energy saving imply boosting technological progress and restructuring the economy and social sector by implementing economic, administrative, and regulatory approaches to increase energy efficiency and energy saving. Research shows that the technological transformation of the national economy and social environment is a crucial factor in these spheres [9]. Suppose Ukraine can achieve energy efficiency indicators similar to those in other countries. In that case, it can save hundreds of millions of tons of waste fuel owing to technological improvements. The general transformation of the national economy, industry, and business through technological modernization includes decommissioning outdated equipment, gradual withdrawal of energy-inefficient goods, and introducing the most recent technologies, equipment, meters, and systems.

3   Results and Discussion

3.1   Infrastructural and Marketing Support of the Energy-Saving System

The strategic direction for developing the infrastructure and marketing support of the energy-saving system involves applying a systematic approach to developing and promoting energy-saving programs to achieve maximum efficiency and sustainability of projects. The proposed scientific and methodological approach to the infrastructure and marketing support of the energy-saving system at the first stage of its implementation includes a detailed analysis of energy efficiency, the condition of buildings available, and their infrastructure and heating systems. The results of such an analysis are the basis for the technical and economic reasoning of projects on energy efficiency (Fig. 1).

The results obtained by assessing the effectiveness of energy-saving measures and technologies, based on their technical characteristics and cost efficiency, allow one to create a basis for involving newly motivated participants to encourage the increase in energy efficiency of buildings and scale up positive experiences of implementing energy-saving projects.

Building owners, enterprises, state bodies, and public organizations can act as the parties concerned. Currently, the only way to encourage people to participate in energy-saving programs is by providing infrastructure and marketing support. This includes developing strategies and tools such as information campaigns, websites, seminars, and other activities to increase public awareness and motivate participants.

It is also necessary to provide tax subsidies and incentives to support energy-efficiency projects, which may include tax reductions for facilities implementing energy-saving technologies or providing financial privileges to investors in energy efficiency. The proposed approach makes it possible to systematically resolve existing issues in the national energy sector and, based on monitoring the achieved results, develop effective strategies for further development aimed at constantly increasing the level of energy efficiency, sustainability and security of the economy, and scaling up energy efficiency programs by the Project of the Recovery Plan of Ukraine ("Energy Security") [3].

The primary focus of recovery programs should be on improving energy efficiency and implementing energy-saving tasks in infrastructure facilities within the postal communication branches of "Ukrposhta" JSC. This provider of postal services in Ukraine plays a vital role in the postal system and is essential for facilitating communication between various regions of the country. Speedy and dependable mail delivery services are offered by them, which are crucial for fostering communication among local communities and regional hubs.

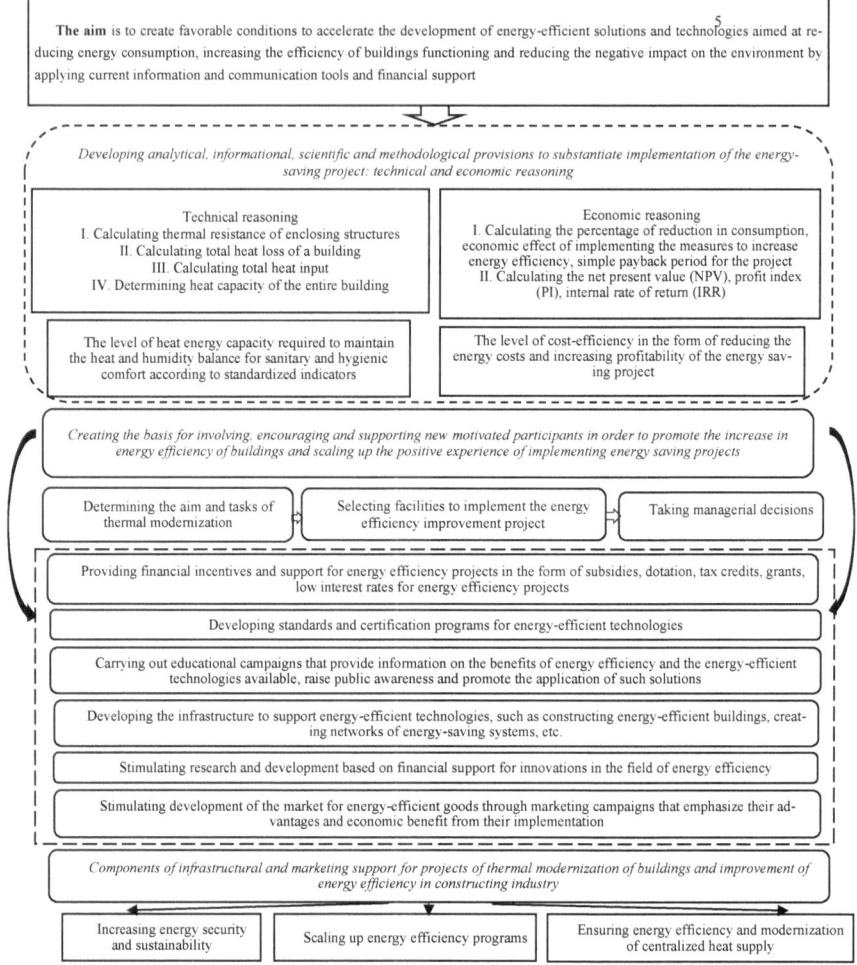

Fig. 1 Scientific and methodological approach to infrastructural and marketing support of the energy-saving system

3.2 Justification of the Feasibility of Implementing an Energy-Saving Project on the Example of National Postal Service Provider

The feasibility of implementing an energy-saving project that can be scaled up within the framework of the national postal service provider's functioning and contribute to developing effective and innovative solutions in the field of energy-saving components of postal service infrastructure and logistics should be justified on the example of Sumy branch of "Ukrposhta" JSC.

The central systems that ensure the operation of the Sumy branch of "Ukrposhta" JSC are heat, electricity, water supply, ventilation, and sewage. Analyzing the energy consumption volumes by types of energy supply systems at the Sumy branch of "Ukrposhta" JSC showed that the branch spends 56% of the total funds dedicated to energy and water supply on heating, and that requires effective solutions regarding the implementation of energy efficiency and energy-saving programs. The energy supply controlling system of "Ukrposhta's" Sumy branch was analyzed on the basis of a normative methodology for calculating the thermal resistance of enclosing structures, which allowed the following conclusions to be drawn [10].

For external enclosing structures of heated buildings and constructions, the following conditions must be satisfied:

$$R_{\Sigma\text{pr}} \geq R_{q\min}, \tag{1}$$

where

$R_{\Sigma\text{pr}}$ is the current thermal resistance of a solid building enclosure or a solid component of a building enclosure, $\frac{\text{m}^2\,\text{K}}{\text{W}}$;

$R_{q\min}$ is the minimum acceptable value of the heat transfer resistance of an opaque enclosing structure or an opaque part of enclosing structure, $\frac{\text{m}^2\,\text{K}}{\text{W}}$.

The heat transfer resistance value must meet a minimum standard for opaque enclosing structures, translucent enclosing structures, doors, and gates in industrial buildings. This minimum value, referred to as $R_{q\min}$, is determined according to [10] and is based on the building's temperature zone, thermal and humid conditions, and the thermal inertia of the enclosing structures D, calculated using a specific formula:

$$D = \sum_{i=1}^{n} R_i \cdot s_i, \tag{2}$$

where

R_i is thermal resistance of the i-th layer of the structure, which is calculated with the help of the formula:

$$R_i = \frac{\delta_i}{\lambda_{i_\text{p}}}, \tag{3}$$

where

δ_i is thickness of the i-th layer of the structure, m;

λ_{i_p} is thermal conductivity of the i-th layer's material of the structure in the calculated operating conditions, W/(m K), taken according to [10];

s_i is coefficient of heat absorption of the i-th layer's material of the structure in the calculated operating conditions, W/(m^2 K), taken according to [10];

When verifying that the criteria outlined in formula (1) are satisfied, the current thermal resistance of the opaque enclosure is determined using the formula:

$$R_{\Sigma n_p} = \frac{1}{\alpha_{out}} + \sum_{i=1}^{n} R_i + \frac{1}{\alpha_{in}} = \frac{1}{\alpha_{out}} + \sum_{i=1}^{n} \frac{\delta_i}{\lambda_{i_p}} + \frac{1}{\alpha_{out}} \tag{4}$$

where

$\alpha_{in}, \alpha_{out}$ are coefficients of heat transfer of inner and outer surfaces of an enclosing structure, $W/(m^2 \ K)$, taken according to [10];

λ_{i_p} is thermal conductivity of the i-th layer's material of the structure in the calculated operating conditions according to [10], $W/(m^2 \ K)$;

n is the number of layers in the structure by direction of the heat flow;

R_i is thermal resistance of the i-th layer of the structure, according to formula (3), $\frac{m^2 \ K}{W}$;

The heat transfer resistance of the material placed in the light slots of windows is determined in accordance with reference [10].

The formula provided below is used to determine the current thermal resistance of the exterior wall:

$$R_{\Sigma pr} = \frac{1}{8.7} + \frac{0.38}{0.81} + \frac{0.03}{0.81} + \frac{0.03}{0.81} + \frac{1}{23} = 0.70 \ \frac{m^2 \ K}{W}, \tag{5}$$

The factual value of thermal resistance is compared to the minimum acceptable one [7]:

$$R_{\Sigma pr} = 0.70 \ \frac{m^2 \ K}{W} < R_{q_{min}} = 3.3 \ \frac{m^2 \ K}{W}, \tag{6}$$

The values of the present heat transfer resistance are much smaller than of the standardized one, so it can be said that the walls are not sufficiently insulated. The present thermal resistance of the ceiling is calculated in a similar way:

$$R_{\Sigma pr} = \frac{1}{8.7} + \frac{0.23}{2.04} + \frac{0.015}{0.12} + \frac{0.004}{0.17} + \frac{1}{23} = 0.41 \ \frac{m^2 \ K}{W}, \tag{7}$$

The factual value of thermal resistance is compared to the standardized one [7]:

$$R_{\Sigma pr} = 0.41 \ \frac{m^2 \ K}{W} < R_{q_{min}} = 4.95 \ \frac{m^2 \ K}{W}, \tag{8}$$

Based on the results of calculating the present heat transfer resistance of the ceiling, it can also be concluded that the roof needs insulation. When calculating thermal resistance of plastic doors, $R_{\Sigma pr}$ value is taken according to the table [6]:

$$R_{\sum \text{pr}} = 0.4 \ \frac{\text{m}^2 \ \text{K}}{\text{W}}, \tag{9}$$

For plastic windows, $R_{\sum \text{pr}}$ value is taken according to table [6]:

$$R_{\sum \text{pr}} = 0.54 \ \frac{\text{m}^2 \ \text{K}}{\text{W}}, \tag{10}$$

The present thermal resistance of the floor is calculated according to the formula:

$$\sum R_{n_2} = \frac{0.005}{1.1} + \frac{0.22}{0.38} + \frac{0.06}{0.7} = 0.21 \ \frac{\text{m}^2 \ \text{K}}{\text{W}}, \tag{11}$$

Thermal heat transfer resistance of individual floor zones on the ground R_{fg}, $(\text{m}^2 \ {}^\circ\text{C})/\text{W}$ is determined by the formulas:

$$\text{Zone I---}R_{\text{fg}}^{\text{I}} = R_0^{\text{I}} + \sum R_n;$$
$$\text{Zone II---}R_{\text{fg}}^{\text{II}} = R_0^{\text{II}} + \sum R_n;$$
$$\text{Zone III---}R_{\text{fg}}^{\text{III}} = R_0^{\text{III}} + \sum R_n;$$
$$\text{Zone IV---}R_{\text{fg}}^{\text{IV}} = R_0^{\text{IV}} + \sum R_n;$$

where

$R_0^{\text{I}}, R_0^{\text{II}}, R_0^{\text{III}}, R_0^{\text{IV}}$ are the values of thermal heat transfer resistance of individual floor zones on the ground, $(\text{m}^2 \ {}^\circ\text{C})/\text{W}$, which, respectively, equal to 2.2; 4.3; 8.6; 14.2;

$\sum R_n$ is the sum of values of thermal heat transfer resistance of the floor layers on the ground, $(\text{m}^2 \ {}^\circ\text{C})/\text{W}$.

The value of $\sum R_n$ is calculated according to the equation:

$$\sum R_n = \sum_{i=1}^{n} \frac{\delta_i}{\lambda_i}, \tag{12}$$

where

n is the number of floor layers on the ground;
δ_i is thickness of the i-th layer, m;
λ_i is thermal conductivity coefficient of the material's i-th layer, $(\text{m}^2 \ {}^\circ\text{C})/\text{W}$.

Thermal resistance of each zone is calculated according to the following formulas:

$$\text{Zone I---}R_{\sum \text{pr}} = 2.2 + 0.21 = 2.41 \ \frac{\text{m}^2 \ \text{K}}{\text{W}}, \tag{13}$$

$$\text{Zone II---}R_{\textstyle\sum\text{pr}} = 4.3 + 0.21 = 4.51 \ \frac{m^2 \ K}{W}, \tag{14}$$

$$\text{Zone III---}R_{\textstyle\sum\text{pr}} = 8.6 + 0.21 = 8.81 \ \frac{m^2 \ K}{W}, \tag{15}$$

The results of calculating thermal resistance are presented in Table 1.

It can be seen from Table 1 that the values of present heat transfer resistance here are much lower than the valid values, which indicates the need to insulate the enclosing structures, that is, outer walls and ceiling in this particular case. It is necessary then to calculate the heat losses by enclosing the building's structures (walls, windows and doorways, ceilings, and uninsulated floors).

The total heat losses of a building are determined according to the standard methodology [11].

Heat losses through outer walls are calculated with the formula:

$$\sum Q_{\text{wall}} = \frac{1187.91}{0.70} * (20 + 22) * 1 = 71,108.9 \ W, \tag{16}$$

A similar procedure is used to calculate heat losses through the ceiling:

$$\sum Q_{\text{ceil}} = \frac{539.01}{0.41} * (20 + 22) * 1 = 53,940.78 \ W, \tag{17}$$

Heat losses through windows are calculated as follows:

$$\sum Q_{\text{win}} = \frac{2.88 * 79}{0.54} * (20 + 22) = 21,504 \ W, \tag{18}$$

Table 1 Results of heat transfer resistance calculations at Sumy Branch of "Ukrposhta" JSC

Type of enclosing structure		Valid value of heat transfer resistance $R_{q_{\min}}$, m² K/W	Present thermal resistance $R_{\Sigma\text{pr}}$, m² K/W
Outer wall		3.3	0.70
Ceiling		4.95	0.41
Window			
Plastic		0.54	–
Door			
Wooden		–	0.33
Plastic		–	0.4
Floor	Zone I	–	2.41
	Zone II		4.51
	Zone III		8.81
	Zone IV		–

Heat losses through doorways are calculated with the formula:

$$\sum Q_{door} = \frac{4 * 2.6}{0.4} * (20 + 22) = 1092 \text{ W}, \qquad (19)$$

Heat losses through floors are calculated in the following way:

$$\sum Q_{fl} = \left(\frac{205.04}{2.41} + \frac{2173.04}{4.51} + \frac{160.93}{8.81} \right) * (20 - 6) = 1980.51 \text{ W}, \qquad (20)$$

where

$$F_I = (42.61 * 12.65 - 38.61 * 8.65) = 205.04 \text{ m}^2,$$
$$F_{II} = (38.61 * 8.65 - 34.61 * 4.65) = 173.04 \text{ m}^2,$$
$$F_{III} = (34.61 * 4.65) = 160.93 \text{ m}^2.$$

Total heat losses through enclosing structures constitute:

$$\sum Q_0 = 71,108.9 + 53,940.78 + 21,504 + 1092 + 1980.50 = 149,626.24 \text{ W}, \qquad (21)$$

Additional heat losses through external walls due to location of buildings are calculated with the following formula:

$$Q_{loc}^{ad} = 71,108.9 * 0.13 = 9244.16 \text{ W}, \qquad (22)$$

Additional heat losses due to door opening are:

$$Q_{d.op}^{ad} = 1092 * 4 = 4368 \text{ W}, \qquad (23)$$

Additional heat losses due to uninsulated floors and location on the ground are:

$$Q_{fl}^{ad} = 0.05 * 1980.5 = 99.02 \text{ W}, \qquad (24)$$

Additional heat losses along the room height are:

$$Q_h^{ad} = 0.02 * 71,108.9 = 1422.17 \text{ W}, \qquad (25)$$

The total amount of additional heat losses through enclosing structures constitutes:

$$\sum Q^{ad} = 9244.156 + 4368 + 99.02 + 1422.7 = 15,133.36 \text{ W},$$

Additional heat losses due to air infiltration through windows should be calculated with the following formula:

$$Q_{\text{win}}^{\text{inf}} = 0.28 * 0.5 * 276.48 * 1.005 * (20 + 22) = 1633.89 \text{ W}, \qquad (26)$$

Additional heat losses due to air infiltration through doorways are calculated with the formula:

$$Q_{\text{door}}^{\text{inf}} = 0.28 * 196 * 1.005 * (20 + 22) = 2316.48 \text{ W}, \qquad (27)$$

where

$$\begin{aligned}
G_z &= 0.005 * (2 * 2 + 0.8 * 2) * 0.5 * 1.3 * 3600 \\
&\quad + 0.005 * (0.85 * 2 + 2 * 2.1) * 0.5 * 1.3 * 3600 \\
&\quad + 0.005 * (2 * 0.63 + 2 * 2) * 0.5 * 1.3 * 3600 = 196 \text{ kg/h}
\end{aligned}$$

Total additional heat losses due to cold air infiltration are:

$$Q^{\text{ad.inf}} = 1633.89 + 2316.48 \text{ W}.$$

Total estimated heat losses of the premises constitute:

$$\sum Q_{\text{tot}} = 149{,}626.24 + 15{,}133.36 + 3950.31 = 168{,}709.92 \text{ W}, \qquad (28)$$

It is also necessary to analyze the dynamics of heat supply in the studied room. Heat input from people is calculated according to the formula:

$$Q_{\text{p}} = 102 * 110 = 11{,}220 \text{ W}, \qquad (29)$$

Heat input from electrical equipment operating, computers in particular, is:

$$Q_{\text{eq}} = 400 * 110 * (1 - 0.9 * 0.9 + 0.75 * 0.9 * 0.9) * 0.35 = 12{,}281.5 \text{ W}, \qquad (30)$$

Heat input from the lighting sources is:

$$Q_{\text{l}} = 70 * 40 * 0.04 * 0.72 = 806.4 \text{ W}, \qquad (31)$$

Total heat input to the building is:

$$Q_{\text{tot}} = 11{,}220 + 12{,}281.5 + 806.4 = 24{,}307.9 \text{ W}, \qquad (32)$$

So, thermal capacity of the entire building is as follows:

$$\Delta Q = 16{,}879.92 - 24{,}307.9 = 144{,}402.02 \text{ W}. \qquad (33)$$

Energy-saving measures refer to techniques, approaches, or actions taken to decrease energy usage without compromising on comfort or efficiency. These actions could involve the use of energy-saving technologies, improving processes and energy management, and implementing renewable energy sources. The goal of energy-saving initiatives is to decrease energy expenses, lower pollution emissions, conserve natural resources, and minimize energy supply expenses. Ways to conserve energy include:

- free and affordable measures carried out by the company during its normal operations;
- average-cost measures are funded using the internal funds of the enterprises;
- high-cost measures are put in place to draw in external investment.

For modernizing Sumy branch of "Ukrposhta" JSC, energy-saving measures are critically important, as they allow one to reduce energy costs and increase the overall efficiency of resource management. In order to prevent heat from escaping from the national postal service provider's facilities, energy-efficient steps need to be taken. These include upgrading the heating system, adding insulation to the exterior walls, and insulating the roof, all of which will help enhance the buildings' thermal efficiency. All this will lead to reduced heating costs and increased comfort for employees working on the premises of the postal service provider. In order to increase the cost efficiency of the heating station functioning, it is proposed to modernize it by installing KIARM combined regulator 63022, which provides:

- supporting constant differential pressure (spending) in facilities' heat supply system when changing hydraulic parameters in the input and output lines;
- obtaining pressure change parameters (spending) with different air temperatures, periods, and days of the week;
- maintaining thermal comfort in the areas serviced with minimal energy consumption.

The temperature regulator is designed to maintain the set temperature in the area serviced, with automatic correction upon changes in external parameters depending on the periods established.

Considering the amount of thermal energy consumed by Sumy branch of "Ukrposhta" JSC for the year 2023 and the current tariffs, the cost of thermal energy for the year constitutes UAH 328,743.24. To justify the feasibility of implementing the proposed measures, the necessary economic costs should be calculated. The total costs for implementing the project (regulator, controller, sensors, and equipment installation) are UAH 12,108.

Let the cost efficiency of installing the regulator be considered. The coefficient of heat consumption reduction is:

$$r_R = \frac{t \cdot \xi_{R_1} - t_z}{t \cdot \xi_{R_2} - t_z}, \qquad (34)$$

where

t is the temperature inside the building;

t_z is the average ambient temperature during the heating season;

ξ_{R_1}, ξ_{R_2} are quality coefficients of regulating and technical equipment of the system
          consistent with basic and actual variant of the project solution.

Percentage reduction in consumption is:

$$k = (1 - r_R) \cdot 100\%. \tag{35}$$

If $\xi_{R_1} = 0.5$, then there are no thermostats or auto-regulators in the system at
the input point but there is central regulation in the central heating station or boiler
room; when $\xi_{R_2} = 0.7$, there are no thermostats in the system but there is central
auto-regulation at the input point.

Here is the coefficient of heat consumption reduction determined:

$$k = (1 - 0.75) \cdot 100\% = 25\%$$

The cost efficiency substitutes:

$$328{,}743.24 * 0.25 \approx 82{,}185.81 \text{ UAH.}$$

The project's payback period may be calculated as follows:

$$P = 12{,}108/85{,}185.81 = 0.15 \text{ year.}$$

In a similar way, the cost efficiency of insulating the outer walls is calculated. The
present heat transfer resistance, $R_{\Sigma\text{pr}}$, m^2 K/W is calculated with the formula:

$$R_{\Sigma\text{pr}} = \frac{1}{\alpha_{\text{in}}} + \sum_{i=1}^{n} R_i + \frac{1}{\alpha_{\text{out}}} = \frac{1}{\alpha_{\text{in}}} + \sum_{i=1}^{n} \frac{\delta_i}{\lambda_{i_c}} + \frac{1}{\alpha_{\text{out}}}, \tag{36}$$

where

$\alpha_{\text{in}}, \alpha_{\text{out}}$ are heat transfer coefficients of inner and outer surfaces of the enclosing
          structure, W/(m^2 K), which are taken according to [11];

λ_{i_c} is material's thermal conductivity in the i-th layer of the structure in the
          calculated terms of use according to [11], W/(m K);

n is the number of layers in the structure, following the heat flow direction;

R_i is thermal resistance of the i-th layer of the structure.

$$R_{\Sigma\text{pr}} = \frac{1}{8.7} + \frac{0.38}{0.81} + \frac{0.03}{0.81} + \frac{0.03}{0.81} + \frac{1}{23} = 0.70 \text{ m}^2 \text{ K/W.}$$

The factual value of thermal resistance should be compared with the standardized
one according to National Construction Norms B.2.6–31:2006 "Constructions of

buildings and structures. Thermal insulation of buildings" [11]:

$$R_{\Sigma pr} = 0.70 \text{ m}^2 \text{ K/W} < R_{q min} = 3.3 \text{ m}^2 \text{ K/W}.$$

Since the factual resistance value is less than the standardized one, it is recommended to insulate the walls to increase thermal resistance and reduce heat losses. Insulation is carried out with EPS-C 25 foam plastic. Thermal conductivity of a foam plastic plate is 0.039 W/(m °C). The required thickness of δ_{ins} thermal insulation layer for insulating the outer enclosing structure is:

$$\Delta_{ins} = (3.3 - 0.70) * 0.039 = 0.10 \text{ m}.$$

The total cost of insulating outer walls of the building (installation and decoration works included) is UAH 254,194. The new calculated value of heat transfer resistance of the multilayer enclosing structure will be as follows:

$$R_{\Sigma pr} = 0.71 + \frac{0.1}{0.039} = 3.3 \text{ m}^2 \text{ K/W}.$$

The obtained value satisfies the condition $R_{\Sigma av} > R_{q min}$.
Here is the reduction of heat losses of the building due to implementing the energy-saving measures calculated:

$$\Delta Q_0 = 853 * (20 + 22) * 1 * \left(\frac{1}{0.71} - \frac{1}{3.3} \right) = 61,917 \text{ W}.$$

Savings per year will constitute:

$$\Delta Q = 61,917 * 185 * \frac{24 - (-2.5)}{24 - (-22)} * 24 * 0.8598 * 10^{-6} = 136 \text{ W}.$$

The cost efficiency of implementing the energy-saving measures is:

$$E = 717.78 * 136 = \text{UAH } 97,740.$$

Simple payback period for the energy-saving measures is as follows:

$$T = \frac{254,194}{97,740} = 2.6 \text{ years}.$$

The cost efficiency of roof insulation should also be determined below. This is to calculate thermal resistance of the roof:

$$R_{\Sigma prr} = \frac{1}{8.7} + \frac{0.23}{2.04} + \frac{0.015}{0.12} + \frac{0.004}{0.17} + \frac{1}{23} = 0.41 \text{ m}^2 \text{ K/W}.$$

Then, the factual value of thermal resistance is compared with the standardized one according to National Construction Norms B.2.6–31:2006 "Constructions of buildings and structures. Thermal insulation of buildings" [12]:

$$R_{\Sigma pr} = 0.41 \text{ m}^2 \text{ K/W} < R_{q_{min}} = 4.95 \text{ m}^2 \text{ K/W}.$$

Since the factual resistance value is less than the standardized one, it is recommended to insulate the roof to increase thermal resistance and reduce heat losses. Insulation is carried out with EPS-C 25 foam plastic. Thermal conductivity of a foam plastic plate is 0.039 W/(m °C). The required thickness of δ_{ins} thermal insulation layer for insulating the outer enclosing structure is calculated as follows:

$$\delta_{ins} = (4.95 - 0.41) * 0.039 = 0.17 \text{ m}.$$

Then, the closest standard value of the plate thickness available in sale is taken, and that is 0.1, 0.05 and 0.02 m. The total cost of the building's roof insulation is UAH 98,064. The new calculated value of heat transfer resistance of the multilayer enclosing structure will be:

$$R_{\Sigma pr} = 0.41 + \frac{0.17}{0.039} = 4.95 \text{ m}^2 \text{ K/W}.$$

The obtained value satisfies the condition $R_{\Sigma pr} > R_{q_{min}}$. Reduction of heat losses of the building due to implementing the energy-saving measures is as follows:

$$\Delta Q_0 = 540 * (20 + 22) * 1 * \left(\frac{1}{0.41} - \frac{1}{4.95} \right) = 58,514 \text{ W}.$$

Savings per year will constitute:

$$\Delta Q = 58,514 * 185 * \frac{24 - (-2.5)}{24 - (-22)} * 24 * 0.8598 * 10^{-6} = 128 \text{ W}.$$

Here is the cost efficiency of implementing the energy-saving measures calculated:

$$E = 717.78 * 128 = 92,368 \text{ UAH},$$

Along with simple payback period for energy-saving measures:

$$T = 98,064/92,368 = 1 \text{ year}.$$

The proposed measures made it possible to reduce annual costs of thermal energy by UAH 272,290; this is by more than 5.8 times.

3.3 Determining the Economic Efficiency of Implementing an Energy-Saving Measure Using the Discount Method

In addition, the cost efficiency of implementing the energy-saving measures using a discount method should be determined, based on the following conditions: the amount of capital expenditures is UAH 254,194; the project life cycle aimed at cutting the heat energy costs by reducing heat losses through enclosing structures is 40 years; the discount rate is 25%. Net present value is calculated according to the formula [13]:

$$NPV = \sum_{t=0}^{n} \frac{R_t}{(1+i)^t},$$ (37)

where

R_t represents the total amount of cash coming in and going out during a specific time frame;

i is the discount rate or return that can be achieved through other investment options;

t represents the number of time intervals.

As a result of the above calculations, the net present value is obtained at the level of:

$$NPV = 955,804 - 254,194 = 701,610 \text{ UAH}.$$

The result of the NPV calculation is the indicative criterion for deciding to invest in the energy-saving project. In this case, $NPV > 0$ is when the present value of cash inflows exceeds the present value of cash outflows. The project then is profitable and worth undertaking. The calculations also showed that, in absolute terms, the project pays back in 2 years or 4 years when considering the discount rate. The net profit of the project is UAH 3,655,406, while the net present value is UAH 701,610.

The profitability index (PI) is calculated according to the formula:

$$PI = \frac{\text{Present Value (PV) of Future Cash Flows}}{\text{Initial Investment}}$$

$$PI = \frac{955,804}{254,194} = 3.7.$$ (38)

Since $PI > 1$, the present value of cash inflows exceeds the present value of cash outflows. The project then is profitable and worth undertaking. Internal rate of return is understood as the calculated present rate when the profit from project implementation is equal to the present implementation costs. Thus, internal rate of return (IRR) is a discount rate at which $NPV = 0$. When the discount rate increases, the net present value decreases and turns into zero at a certain value of the discount rate. The discount rate at which NPV equals zero is the internal rate of return [14]. The iteration method is used to calculate IRR in Microsoft Excel. The IRR function performs a cyclic calculation until the result approaches the value of assumption argument to the nearest of 0.00001%. The result obtained is 38%. To check the obtained result, NPV with discount rates of 37 and 39% should be calculated. NPV at 37% discount rate constitutes UAH 9967. NPV at 39% discount rate is UAH 3579. IRR > r, that is, IRR exceeds the minimum investment costs for this project, and therefore, the project is worth undertaking.

The discounted payback period is calculated according to the formula:

$$DPP = \sum_{i=1}^{n} \frac{CF_i}{(1+r)^i},$$

(39)

where

CF_i represents the cash flow produced by the investment project;
r is a rate for discounts;
n represents the duration of the project being carried out.

$$DPP = 3 + \frac{254,194 - 243,065}{66,758} = 3.16.$$

The given energy-saving measure is cost efficient, since NPV > 0. Therefore, cash inflows' present value exceeds cash outflows' present value. The project is attractive for investment and can help multiply the enterprise capital and increase its market value, so it is highly probable to implement the project successfully [15]. Similar to the previous calculations, the cost efficiency of modernizing the heating station and insulating the roof is calculated (Table 2).

Table 2 Expected economic indicators of implementing the energy-saving measures in the premises of Sumy Branch of "Ukrposhta" JSC

№ ref.	The list of indicators	Outer walls insulation	Heating station modernization	Roof insulation
1.	Capital investments, UAH	254,194	12,108	98,064
2.	Annual operating costs, UAH	–	–	–
3. Technical and economic indicators				
3.1	Total cost of the goods produced (annual savings), UAH	97,740	82,185	92,367
3.2	Net present value, UAH	701,610	791,583	805,198
3.3	Profitability index	3.7	66.37	9.21
3.4	Internal rate of return, %	38	679	94
3.5	Discounted payback period, years	3.16	0.16	1.18

Thus, all the proposed energy-saving measures are cost efficient since NPV > 0; that is, the present value of cash inflows exceeds the present value of cash outflows. The projects are attractive for investment and can contribute to strengthening energy sustainability and increasing the market value of the Sumy branch of "Ukrposhta" JSC.

Similar projects are worth undertaking at all transport infrastructure facilities of the national postal service provider. This will allow one to achieve the goals set in the Project of the Recovery Plan of Ukraine to achieve energy security by increasing energy efficiency and demand management [6].

4 Conclusions

In the context of the conducted research, substantiating the projects on implementing energy-efficient technologies into the infrastructure network of the national postal service provider and scaling up this experience to post offices and other premises of "Ukrposhta" JSC creates prerequisites for increasing energy efficiency in the construction industry by 35%. Structural modernization and full integration into the EU, as noted by the "Energy Security" working group members of the Recovery Plan of Ukraine, are possible only through massive thermal modernization of buildings and future construction of nearly zero-energy buildings (NZEB). To achieve the goals set, it is necessary to create an effective marketing and tax support system for the energy-saving system based on strategic guidelines for Ukraine's national economic recovery programs. These are critical components for ensuring the successful implementation of energy efficiency measures at infrastructure facilities.

Marketing strategies form a responsible attitude and raise awareness of citizens and businesses regarding the importance of implementing energy-saving projects among the concerned parties by involving the public and businesses in energy-saving programs, creating demand for energy-efficient technologies and services, and increasing public support for implementing necessary measures. Marketing support of the energy-saving system allows one to draw attention and find resources to resolve the relevant issues in the field of energy efficiency, stimulate the implementation of energy-efficient solutions, and promote cooperation among various parties concerned.

Tax instruments may also be applied to stimulate the implementation of energy-efficient measures by providing tax benefits, reducing taxes on energy-saving equipment imports, implementing projects on the thermal modernization of existing buildings, or constructing new modern, nearly zero-energy buildings. Tax support for stimulating investment in energy-efficient technologies and infrastructure, together with reducing financial barriers to implementing energy-saving programs, also creates prerequisites for active development and well-being of the energy-efficient solutions market in the process of restoring the national economy and ensuring sustainable growth on a long-term basis. Financial incentives generally cover two main areas:

saving traditional energy resources and developing alternative energy sources. Directions for further research include the development, implementation, and marketing promotion of energy-saving projects to achieve the strategic goals of national economic recovery.

References

1. M.B. Pushkareva, G.V. Tikhonov, M.D. Pushkarev, Improving the energy efficiency of industrial buildings. Russ. Eng. Res. **44**, 466–468 (2024). https://doi.org/10.3103/S1068798X24700199
2. S. Momeni, A. Eghbalian, M. Talebzadeh et al., Enhancing office building energy efficiency: neural network-based prediction of energy consumption. J. Build. Rehabil. **9**, 68 (2024). https://doi.org/10.1007/s41024-024-00416-4
3. F.S. Hafez, B. Sa'di, M. Safa-Gamal, Y.H. Taufiq-Yap, M. Alrifaey, M. Seyedmahmoudian, S. Mekhilef, Energy efficiency in sustainable buildings: a systematic review with taxonomy, challenges, motivations, methodological aspects, recommendations, and pathways for future research. Energy Strategy Rev. (2023, Jan 1). https://doi.org/10.1016/j.esr.2022.101013
4. J. Billanes, P. Enevoldsen, Influential factors to residential building occupants' acceptance and adoption of smart energy technologies in Denmark. Energy Build. **276** (2022). https://doi.org/10.1016/j.enbuild.2022.112524
5. State Agency for Energy Efficiency and Energy Saving, National Energy Efficiency Action Plan until 2030. https://saee.gov.ua/uk/content/npdee-2030
6. Ministry of Energy of Ukraine, Energy Strategy of Ukraine until 2050. https://www.mev.gov.ua/reforma/enerhetychna-stratehiya. ЗаконУкраїни Про внесення змін до деяких законів України щодо запобігання зловживанням на оптових енергетичних ринках. URL: https://zakon.rada.gov.ua/laws/show/3141-20#Tex
7. Cabinet of Ministers of Ukraine, Recovery Plan of Ukraine ("Energy Security"). https://ua.urc-international.com/plan-vidnovlennya-ukrayini
8. Cabinet of Ministers of Ukraine on Energy Efficiency of Buildings Law of Ukraine. https://zakon.rada.gov.ua/laws/show/2118-19#Text
9. N. Kuzmynchuk, O. Terovanesova, T. Kutsenko, O. Zyma, I. Bachkir, Paradigm towards ensuring of energy saving in the crisis management conditions in the aspect of sustainable environmental development, in *International Conference on Sustainable, Circular Management and Environmental Engineering (ISCMEE 2021)*. E3S Web Conference, vol. 255, 5, 01022 (2021). https://doi.org/10.1051/e3sconf/202125501022
10. Ministry of Regional Development, Construction and Housing and Communal Services of Ukraine on Approval of the Methodology for Determining the Energy Efficiency of Buildings Order. https://zakon.rada.gov.ua/laws/show/z0822-18#Text
11. National Standard of Ukraine DSTU 9190:2022, Energy Efficiency of Buildings. Method for Calculating Energy Consumption in the Process of Heating, Cooling, Ventilation, Lighting and Hot Water Supply
12. National Construction Norms B.2.6–31:2006, Constructions of Buildings and Structures. Thermal Insulation of Buildings
13. C. Rodriguez-Sanchez, The role of social marketing in achieving the planet sustainable development goals (SDGs). Int. Rev. Public Nonprofit Mark. **20**, 559–571 (2023). https://doi.org/10.1007/s12208-023-00385-3
14. R. Wang, T. Cao, X. He et al., Energy financing, energy projects retrofit and energy poverty: a scenario-analysis approach for energy project cost estimation and energy price determination. Environ. Sci. Pollut. Res. **30**, 108865–108877 (2023). https://doi.org/10.1007/s11356-023-29822-w

15. R. Schmidt-Scheele, W. Hauser, O. Scheel et al., Sustainability assessments of energy scenarios: citizens' preferences for and assessments of sustainability indicators. Energy Sustain. Soc. **12**, 41 (2022). https://doi.org/10.1186/s13705-022-00366-0

Analysis of Financial Stability and Sustainability Approaches in the Energy Sector

Dinara Mukhiyayeva⑩**, Aliya Shakharova**⑩**, Lutpulla Omarbakiyev**⑩**, Aziz Akhmetov**⑩**, and Aigul Alibekova**⑩

Abstract The study investigates the issues with financial stability in the energy sector, focusing on the recommendations for enhancing unstable aspects and the factors that affect stability. This sector has been found to be facing several financial challenges including regulatory issues, market volatility, and technological shocks. These problems have also led to instability of the financial environment, thus making the sector to struggle in search of capital and funding for its operations. To address these vulnerabilities, several measures have been suggested, which include adopting new technology, establishing ways of managing risks, and improving the relations between industries and regulatory bodies. From the regression analysis of computational assessment of the financial contingency of the energy segment, GDP (0.2848), government subsidies (0.1373), and interest rates (0.1916) make the main effects observed. For instance, higher energy price volatility ($-$ 0.0297) is associated with a poor financial stability position which in turn means that tools of risk management are necessary to counter policy changes in energy prices.

Keywords Financial stability · Energy sector · Risk management · Regulatory frameworks · Sustainability

D. Mukhiyayeva · A. Shakharova (✉) · A. Alibekova
L.N. Gumilyov Eurasian National University, 2 Satbaeva Street, 010000 Astana, Republic of Kazakhstan
e-mail: ShaharovaAliya@yandex.kz

L. Omarbakiyev
Turan University, 16A Satpayeva Street, 050013 Almaty, Republic of Kazakhstan

A. Akhmetov
Bank RBK JSC, 15 Republic Square, 050000 Almaty, Republic of Kazakhstan

© The Author(s), under exclusive license to Springer Nature Switzerland AG 2024   81
V. Koval (ed.), *Renewables in the Circular Economy and Business*,
SpringerBriefs in Applied Sciences and Technology,
https://doi.org/10.1007/978-3-031-72174-8_5

1 Introduction

Energy industry is one of the basic sectors of the world economy, which serves as a foundation for manufacturing enterprises, carriage, and households. It is possibly or rather probably the most significant determinant of the growth of the economy because it provides the necessary energy source that runs factories and businesses [1]. This sector is not only limited to satisfying the energy requirement needs. This is because it has various applications in the advancement of technology, creation of employment, and in the process of contributing to the economic system.

In the industrial activities, the energy sector sees to it that various industrial processes are well enhanced and on the go. Thus, there is a stable energy source, which is important for industries to have a production plan schedule on their calendar because they have to supply the market and be competitive [2]. For example, the steel and aluminum industries need a steady source of energy to power several big machines and heaters. Energy disturbances are expensive in terms of financial impacts and the interference with the management of other services in a business environment.

Physical infrastructural support entails constant energy in the movement of commodities and people over long distances through services [3]. For instance, the aviation industry that uses jet fuel has to ensure consistency of supply of this commodity while electricity vehicles require charging stations which requires electricity.

Another important segment of the energy sector formulated based on the importance of residential consumption can also be seen [4]. Energy is one of the most fundamental needs of every home for providing heat and cooling, cooking, and running other electronic appliances that are necessary in the house. On the other side of the equation, the energy using pattern is also shifting in many regions of the world and renewable and sustainable energy sources are being given higher importance because of the change in the environmental and economic conditions [5].

In addition to these direct effects, the energy sector is a major multiplier that creates technological demand and drives innovation [6]. Technologies enable better management of energy and carbon emissions and catalyze the development of new industries and employment domains [7]. For example, solar panel production and usage have resulted in a new industry that deals with the production of panels and installation and maintenance services [6].

Energy sustainability has serious implications for economic security and financial stability in this sector [1]. A common theme that comes into play is the conflict of power over energy resources, affecting entangled trading business [8]. The following are some of the weaknesses of countries that rely on energy imports: countries that rely on imported energy are exposed to world market trends and power battlegrounds because sudden changes in prices and energy accessibility are unpreventable. The energy sector is fundamental to the national economy and its sectors [9]. The topic of this study is the stability of the financial situation in the energy sector, and the following reasons explain its significance. First, it guarantees steady and secure energy unemployment, which is critical in today's global economy.

Sustainability structures increase confidence in financial markets, thus mobilizing the resources crucial for the development of energy projects [10]. They can also make it easier for firms to access capital at low levels, making it easier for them to undertake big ventures such as establishing power generation or renewable energy sources. This is especially the case at this time in the world, when a shift toward cleaner energy sources to meet the critical sustainability agendas is innovation-intensive.

The stability of the energy component in the financial domain is beneficial for increased economic security in the long run as fluctuations in energy prices occur. These risks can be avoided by a financially strong energy sector that provides an economic base for setting up reasonable business situations [8]. For this reason, any strategy designed to mitigate the risks and shape the future of the sector must take cognizance of the challenges outlined above as key to understanding future financial sustainability of how the sector can support sustainable economic development now and in the future.

This study aims to clearly understand the energy sector's soundness prospects as well as future opportunities and threats by discussing the various factors and policies that can influence its financial health and the subsequent strategies that can be implemented to promote stable financial performance in the energy sector. In doing so, the article briefly explains how financial stability in the energy sector is essential for policymakers, industries, and investors and the measures that must be taken to ensure this. The final aim, therefore, is to add to the current discourse on the way forward in enhancing the stability of the energy industry to be fit for the purpose of nurturing economic development and attaining environmental benchmarks.

This study presents a coherent outline to manage the analyses and issues of the financial condition of the energy sector. An introduction explaining why this sector is special and how financial sustainability needs to be addressed. What follows is a literature review section based on published research, theoretical models, and current issues causing instability in the financial sector of the energy industry. The methodology section provides information on the research base based on financial reports of energy companies, industry reports, and reports from regulatory authorities. Results of this study constitutes an analysis and discussion section divided into four key parts: the energy sector's financial performance over the past decade, an analysis of emerging risks and impacts, an assessment of the sector's financial policies and regulations, and ways to create a sustainable financial environment. The discussion and conclusions include a summary of the findings, potential benefits for the sector and the economy, and recommendations for further analysis.

2 Literature Review

The literature on financial stability in the energy sector encompasses a wide array of studies that delve into renewable energy, risk management, regulatory frameworks, and technological innovations. These references provide a comprehensive foundation for understanding the multifaceted dynamics that influence the financial stability of the energy sector.

The literature review of financial stability in the energy sector includes numerous papers that cover renewable energy, risk management, regulation, and technology. These references give a good background from which one can have an overview of the various factors that affect the financial health of the energy industry.

Muller looks into the hydrothermal liquefaction of wasted coffee grounds and their biocatalytic upgrading for biofuel generation with the help of circular economy principles [2]. It is crucial for the current analysis focusing on the opportunities and risks of biofuel production innovations within the energy sector. Thus, applying circular economy concepts, this work identifies the possible economic strategies that may contribute to the improvement of the financial situation.

In this research, Bhola performs an economically and technologically feasible study on the use of rooftops within campus buildings for the generation of solar PV power [11].

Gielen focuses on the place of renewable energy in the global energy transition and underlines the significance of the role of RE in the SDGs [12]. This research work also gives a general idea about the contribution of renewable energy sources in improving the financial stability by minimizing the risk of relying on oil and gas prices and fluctuation in energy prices. The global perspective that is evident in Gielen's study is vital in appreciating the strategic nature of investment in renewable energy.

Mukherjee explores the viability of capturing energy from thunderstorm and comes up with a new source of renewable energy [13].

Prokopenko et al. analyzed how green entrepreneurship is capable of supporting financial stability by encouraging environmental responsible business management and generating new green jobs [10]. Social impact as an element of the research also sheds light on the general economic and environmental effect of adopting green energy measures.

Mazur et al. analyzed the efficient use of capital and risk reduction [9]. This research is directly related to the assessments of the impacts of investments in innovation and technology on the financial results of the energy industry. The study therefore reveals that there is need for organizations to maintain competitive advantage and financial health through innovation.

Trypolska et al. reveal the possibilities of managing the wind and solar power plant end-of-life equipment in Ukraine [4]. This study gives understanding of the effects of managing renewable energy infrastructure in terms of finance and the environment. This paper's analysis of end-of-life management is significant for the analysis of long-term financial and endurance strategies in the energy industry. Kurbatova et al. explore the costs, impacts, and 'brand value' of solar energy for green universities in the case study [1]. This work aims at providing a case study analysis of the costs and advantages of implementing solar energy in institutions. The findings of this study can therefore be employed to elaborate on the impact of RE on FS.

Dovhan, Mokhonyko, and Malyk give a detailed account of the principles of project management that is important in managing large energy projects [14]. The textbook provides information on project management approaches that can contribute to the improvement of the energy sector's operations and financial health.

Thus, the application of these principles will assist energy companies in the effective monitoring of project time, costs, as well as resources, hence reducing on the financial risks that are likely to be incurred. These methods used in this paper have provided a way of properly analyzing the financial indicators and aid in the formulation of sound financial models [15].

Petlenko and Schehlyuk focus on the peculiarities of military technology marketing and offer a novel view on the marketing approaches applicable to the energy industry [5]. Knowing these strategies can assist energy corporate organizations to market their new technologies effectively and enhance their financial accomplishment through reaching more consumers and getting investment. Sopronenkov et al. study the effects of the tax system on the business and economic growth [16]. Thus, the authors show how tax policies can impact the financial situation of energy companies based on the aspects of profit, investment, and the general environment. This paper is important for the analysis of the general economic conditions and legal framework within which energy companies' function.

Timoshenko et al. have explored the integration of Industry 4.0 technologies in organizations, such as automation and data analytics, that can improve the ability of energy companies to manage financial risk and improve productivity. [17].

Kuczabski et al. consider the effectiveness of the regional development management, and therefore present the possibilities of the regional policies and development strategies for the energy sector financial sustainability [18]. Thus, efficient regional management can contribute to the creation of conditions that would be conducive to energy projects, investments, and, consequently, economic development [8]. The study shows how best practices and strategic plans that can be used in improving the financial stability can be implemented.

Masyk et al. outline the criteria of governance's institutional efficiency and quality as applied to sustainable development [7]. The effective governance can enhance the decision-making, risk management and accountability within the companies especially the energy firms that are vital for their financial status.

Bashynska et al. analyze the investment and innovation image of Ukraine's regions in the context of sustainable development [3]. They show that regional investment climates and innovation endowments affect financial stability. Thus, the present work offers an approach to the assessment and enhancement of the investment appeal of regions in the energy sector. Bashynska et al. put forward a conceptual foundation for the development of smart eco-industrial parks as a reference for sustainable production [19]. This research is therefore useful to explain how smart eco-industrial parks can contribute to the creation of financial resiliency through the encouragement of sustainability and innovation in the energy sector.

Dudek et al. analyze the possibilities of inclusive and sustainable development to increase the investment attractiveness of energy companies [20].

The results of these research will be used to generate policy and business recommendations that can help enhance the energy sectors' preparedness and sustainable structure.

3   Methodology

Based on the research objectives, this study employs a quantitative research approach to determine the financial health of the energy sector. The data collected for the research has been gathered from financial statements, industry reports, and other pertinent sources. This paper aims at investigating financial ratios, legal requirements, and market conditions as the factors that affect financial sustainability in the energy industry.

For the analytical approach, both qualitative and quantitative analysis is used. Qualitative analysis is applied to analyze various regulations and guidelines, markets, and industries. Quantitative analysis refers to the various methods of data analysis that are used in finance, these include ratio analysis, trend analysis, and comparative studies. The third area of research used in this study is econometric modeling which is used to assess the effects of various factors on financial stability, and regression analysis to test hypothesis and determine the coefficient of the variables.

The methods employed are fitting to the research objectives and the data collected in the study. Through the use of quantitative analysis, the assessment of the financial indicators of the sector can be done in a more structured and non-biased manner to determine the financial health of the sector. The use of secondary data makes the study reliable and valid since the information used is gotten from other previous works.

4   Results

4.1   Econometric Model for Financial Stability of the Energy Sector

The economic viability of the energy sector is one of the major indicators that defines the readiness and capacity of the sector to fund a sustainable development and meet the enhanced energy needs in society [1]. Overall this will determine its financial health and as a result its ability to support infrastructure, innovations, and market volatility. This is why it is advisable to investigate the motives that lead to financial stability in the energy industry taking into account the trends and contexts prevailing in the contemporary global environment.

The model seeks to bring a certain formulation of what is in fact a complex phenomenon—which is financial stability—and considers how these various indicators contribute positively or negatively to the status of the financial sector.

Study aims at raising awareness of the key factors that should be adequately controlled in order to build and ensuring the stability of the sector among policymakers, industry members, and investors. The model covers a wider scope of factors, which range from the fluctuating energy prices, the regulatory measures, availability of subsidies from the government, macro-economic factors such as gross domestic product and interests' rates, and adoption of technology.

The dependent variable in this model is the financial stability index, measured as FSI which is designed based on the model that create the 'principal component analysis' (PCA) method to analyze and combine several financial indexes into one index. The above cumulative index gives a measure of energy sector's financial viability that involves matters to do with profitability, liquidity, solvency, and efficiency.

The econometric model uses multiple linear regression model to estimate the functional coefficients relating the FSI to its causes. It enables one to gage the manner in which each implied factor as well as their overall cumulative effect on the financial sustenance of the energy sector. The advantage of the proposed model is that the estimates of the weights in ordinary least squares (OLS) are more efficient and less bias than other estimating techniques where total variance is given and fixed, provided the assumptions of the regression analysis have been met.

The facts gathered using this model will add to existing knowledge about the state of financial stability within the energy sector and provide practical information that can help in the formulation of policies, long-term planning, and determination of investments. They are beneficial in the current and future contexts given that the technological advancement of the sector and the changing regulatory environments make it difficult to determine the amount of cash the sector should have on hand in the near and distant future for optimally financing its operations and growth.

The objective of the econometric model is to develop a link between the theories and practicality of this sector, create a platform for policymakers and financial specialists in decision-making, and implement further measures to stabilize the financial state of this sector.

The model will help quantify the relationships between financial stability and various influencing factors such as market risks, regulatory policies, and economic conditions (Table 1).

This research affords an econometric approach for the study of the financial stability of energy sector. Thus, the model can contribute to the improvement of the sector's financial stability by identifying the quantitative links between stability and its determinants.

To this end, regression analysis performed in Stata yields important information about the determinants of the energy sector's financial health. The model analyzes the effects of the energy price fluctuation, regulatory policies, government support, macroeconomic parameters, and technological aspects on the FSI of 200 companies from 15 countries namely USA, Canada, Germany, France, UK, China, Japan, India, Brazil, Russia, Australia, South Korea, Italy, Spain, and South Africa in the years from 2019 to 2023 [21]. The selected countries are 15 in number and were chosen in order to provide a cross section of the global energy market with respect to geography and economic status. This diversification guarantees that all the factors that might affect the energy sector are considered, including the regional policies, market conditions and operations risks, and their effects on the financial stability. Figures 1 and 2 show the figures of the renewable energy capacity in gigawatts, total energy consumption in terawatt-hours, and energy sector growth rate in percentage for each country.

Table 1 Econometric model for financial stability of the energy sector

Dependent variable

FSI

It can be compiled with the help of such factors as profitability, liquidity, solvency, and operating indicators, assessing the financial position of energy companies. By employing PCA to extract the first principal component, it is possible to offer a composite index that would incorporate all the above-listed indicators of financial stability

Independent variables

1. *Market risk factors* 1.1. Energy price volatility (EPV) is defined as the variability of the price of energy in a given period of time expressed as the standard deviation of the prices of energy 1.2. The variant of demand–supply ratio that will be used in this work is the demand–supply imbalance (DSI) or the extent to which supply exceeds or falls short of the demand, expressed as a fraction of the total energy consumed	2. *Regulatory and policy factors* 2.1. RSI stands for Regulatory Stringency Index—a 0–1 representative biodiversity index that encompasses the aspects of energy sector regulations as well as environmental and safety rules and accounting standards 2.2. It encompasses basic grant (BG)—measures for the energy sector excluding subsidies for a specific power source; and total government subsidies (GS)—total subsidies provided to the energy sector as a share of the sector's total revenues
3. *Macroeconomic factors* 3.1. Gross domestic product (GDP)—the total Gross National Product (GNP) or the sum total of all actual and implied economic activity, capable of affecting energy consumption and investment 3.2. Cost of capital—primarily cost of borrowing, or interest rates (IR) which are regulated through central banks and, therefore, have an impact on energy companies' capital costs 3.3. Exchange rates (ER)—this measures the variation in the price of currencies of oil importing countries against the price of currencies of the oil exporting countries	4. *Technological factors* 4.1. Innovation index II/II, a score of technology and innovation of varying energy such as number of patents or R&R of energy 4.2. Renewable energy penetration (REP)-the ratio between renewable energy supplied and the total supply of energy

Model specification

(continued)

Table 1 (continued)

The proposed model is a multiple linear regression model that can be specified as follows:

$$\text{FSI}_t = \beta_0 + \beta_1 \text{EPV}_t + \beta_2 \text{DSI}_t + \beta_3 \text{RSI}_t + \beta_4 \text{GS}_t + \beta_5 \text{GDP}_t \quad (1)$$
$$+ \beta_6 \text{IR}_t + \beta_7 \text{ER}_t + \beta_8 \text{II}_t + \beta_9 \text{REP}_t + \epsilon_t$$

where β_0 is the intercept; $\beta_1, \beta_2, ..., \beta_9$ are the coefficients for each independent variable; ϵ_t is the error term

Estimation method

The model can be estimated using OLS regression, which provides the best linear unbiased estimates of the coefficients, assuming that the Gauss–Markov assumptions hold

Steps for model implementation

1. Data collection (gather data on the dependent and independent variables from reliable sources such as government reports, industry databases, and financial statements of energy companies)	2. Variable construction (construct the FSI using PCA and normalize other variables as necessary)
3. Exploratory analysis (use means and standard deviations as well as correlation analysis in order to determine the nature of the relations between different variables)	4. Model estimation (specifically use statistical software (Stata) to estimate the model using ordinary least square regression)
5. Diagnostic tests (conducted diagnostic tests so as to check for presence of multicollinearity, heteroscedasticity, autocorrelation, and model misspecification errors)	6. Interpretation (explain what the estimated coefficients mean with respect to the various factors that may affect financial stability and the policy recommendations that can be made)

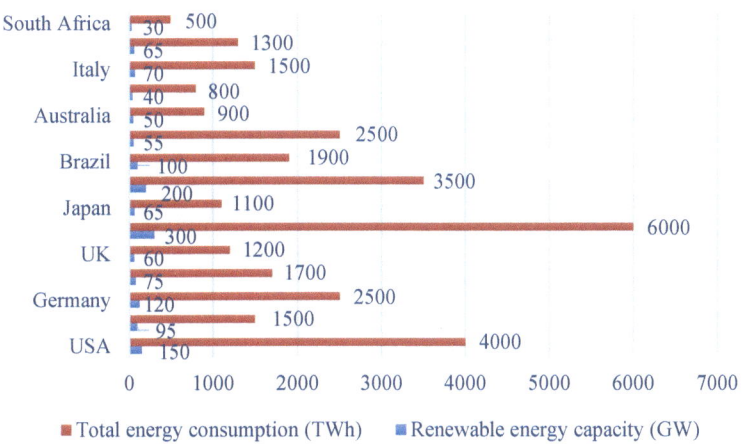

Fig. 1 Energy sector development data for 15 countries years for the period from 2019 to 2023.
Source Based on [22]

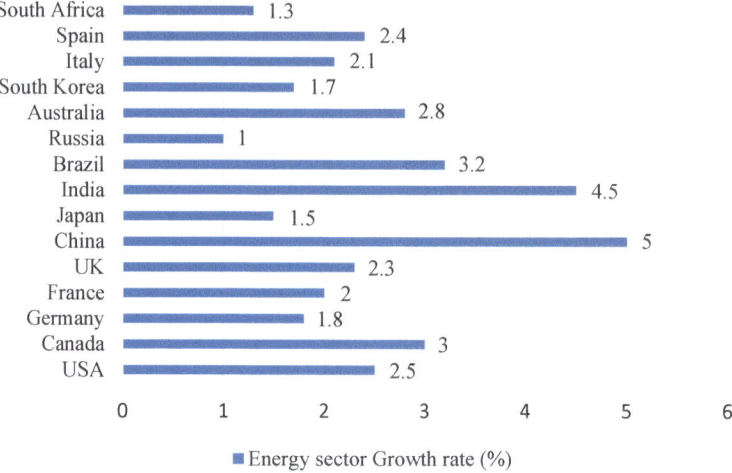

Fig. 2 Energy sector growth rate (percentage) for 15 countries for the period from 2019 to 2023. *Source* Based on [22]

The findings provide a complete picture of the factors influencing the sector's financial health, shining light on important drivers of stability and possible areas for improvement (Fig. 3, Table 2).

Source |	SS	df	MS		Number of obs =	15000
					F(9, 14990) =	3833.29
Model |	3016.5462	9	335.171801		Prob > F =	0.0000
Residual |	670.547276	14990	.044758191		R-squared =	0.6963
					Adj R-squared =	0.6962
Total |	3687.09348	15000	.245806232		Root MSE =	.21165

FSI |	Coef.	Std. Err.	t	P>|t|	[95% Conf.	Interval]
EPV |	-.0297353	.0029151	-10.20	0.000	-.0354601	-.0240106
DSI |	.0123672	.0018474	6.69	0.000	.0087377	.0169968
RSI |	.0524094	.0034176	15.34	0.000	.0457176	.0591012
GS |	.1372695	.0097542	14.07	0.000	.1181494	.1563896
GDP |	.2847652	.0178499	15.95	0.000	.2497286	.3198019
IR |	.1915861	.0256983	7.46	0.000	.1412753	.2418969
ER |	.0043643	.0028542	1.53	0.125	-.0012219	.0099505
II |	.0831464	.0048932	16.99	0.000	.0736163	.0926765
REP |	.0428144	.0046227	9.26	0.000	.0337483	.0518805
_cons |	.5894652	.0263429	22.37	0.000	.5378108	.6411195

Fig. 3 Stata code for regression analysis. *Source* Authors development using Stata program

Table 2 Interpretation of the results from regression analysis and recommendations

№	Factor	Coefficient	Interpretation	Recommendation
1.	EPV	− 0.0297	In turn, an increase in energy price fluctuations has a negative relationship that impacts the level of financial stability	Apply management procedures when facing the danger of varying prices like hedging or diversifying energy supplies
2.	RSI	0.0524	Where regulations are rigorous, more stability is accorded to the financial sector	Pursue strict measures in order to prevent corruption in the energy sector and increase the level of legislation openness
3.	GS	0.1373	Subsidy is one effective way through which the government can come to the assistance of a given entity in enhancing its financial position	Therefore, it is advisable to prescribe subsides for the energy sector especially during regeneration period
4.	GDP	0.2848	Economic growth supports financial stability	Support measures that will enhance economic development
5.	IR	0.1916	Reduced interest rates lead to maintenance of financial integrity	Sustain a low interest rate level that will improve financial a stability in the energy sector
6.	II	0.0831	All the above points make it clear that innovation is beneficial to making good financial provisions	Sustain the expansion of investment in innovation in order to enhance the sectors' readiness to the market shocks
7.	REP	0.0428	As noted above, adoption of renewable energy sources stabilizes the economies	Increase the usage of renewable resources in the generation of electricity, in order to make the sector including electricity prices less dependent on market factors

Source Authors development

EPV has a negative coefficient and asserts the need to control price risks to increase stability. RSI has a positive coefficient, which implies that with the increase in the level of regulation, there is increased financial stability through increased accountability and disclosure.

GS also have a positive effect on the financial stability, so it is possible to state that targeted subsidies improve the situation in the sector. M1 and M2 are positively signed, thus suggesting that economic activities such as GDP and low IR enhance financial stability. Among the technological factors, the indices of II and REP are significant and have positive signs, which means that innovation and the use of renewable energy can help improve financial stability.

The results of the regression analysis are thus useful to policymakers, energy sector actors, and potential investors. Energy price risks, proper regulation, technological innovation, and renewable energy are very important in the enhancement of financial stability. In addition, standing up the economy and offering specific government incentives may contribute to the stability. Thus, tackling these factors, the stakeholders can contribute to the development of the energy sector that would be more resistant to future threats and challenges.

5 Discussion

5.1 Financial Health of the Energy Sector

From the perspective of the concept, renewable energy companies could be more profitable than conventional energy companies due to higher margins resulting from subsidies and the population's focus on clean energy sources [5]. There may be lighter on how these variables influence the profitability by segment in the analysis.

SMEs in the energy sector may experience liquidity problems because they are likely to have limited access to capital than the large corporations [1].

Solvency analysis that examines the capacity of the firms to fulfill long-term commitments can also help to identify different sectors' financial sustainability [8]. This could mean that renewable energy firms exhibit higher solvency ratios because they are more likely to have strategies that will enable them be stable for long than nonrenewable energy companies who may come across some operations risks.

Therefore, by comparing the financial performance of organizations belonging to the various segments in the energy industry, then the stakeholders will be able to understand some of the peculiarities that each segment has to go through or the kind of opportunities that await the particular segment.

Regarding the external factors where financial environment claims its stakes, economic conditions, regulation change, and technology updates are identified priority aspects that affect the energy financial sustainability. The national and global economic status and outlook such as GDP trends, inflation rates, and employment opportunities are factors likely to have an impact on energy utilization and costs. When the economic condition becomes unfavorable, consumers may reduce the utilization of energy and this in turn affects the revenue collect from energy companies and this may adversely affect the companies. On the other hand, some new economic activities require energy which might, in turn, bring about new revenues and a stable financial base.

The changes in the regulative framework are also associated with significant effects on the financial stability of the energy sector [17]. Environmental laws, taxes, and subsidies are some of the factors that influence the business environment of energy companies and their costs. For instance, better and stringent emissions standards may call for expenditure on new and expensive technologies to meet the standards and thus affect the performance of the company financially. Likewise, shifts in taxes and subsidies are key drivers of competition and feasibility in the RE projects. Technological changes, for instance enhanced renewable energy technologies or the emergence of new methods of obtaining fossil fuels, affect the energy sector and its financial health [7].

For instance, the development of renewable energy sources has been boosted by improvements in technology and reduced costs, thus pressuring the conventional fossil fuel industries. This change has forced the conventional energy companies to metamorphose their existing business strategies in order to survive, and this might affect their financial position in the short run.

All these external factors therefore need to be continuously watched by policy makers, regulators, and other stakeholders in the energy sector to avoid future adverse effects on the energy sector.

5.2 Risk Factors and Vulnerabilities

Operational risks such as equipment breakdowns, supply chain issues, and accidents can result in added expenditures and even harm the company's image. Some operational risks may include changes in the regulatory environment, including policies and regulations, which are likely to affect the costs of the energy business and market structure. In evaluating these risks on financial stability, meaning the likelihood of each of them and the level of possible losses in case of their occurrence should be identified.

An example is a situation whereby the price of commodities increases and affects the energy costs trimming down the profits of the firms. Likewise, constraints, such as a strengthened rules and regulations to increase stringency on emissions limits, will relate to increased investments in enhancing technology, which is detrimental on firms' balance sheets. This could lead to product attrition and diminished revenues due to factors such as a fire, storm or any resulting operational risk including a failure in some of the major equipment needed to produce the products [1].

Overcoming these risks requires critical understanding of the interaction between these risks in order to create applicable risk management techniques. For instance, it is possible to lock the price of raw materials to minimize the effect of fluctuations in the price of the commodities or to adopt means and ways of avoiding the losses accruing from operational risks such as those arising from natural disasters. Some of the regulatory risks may mean that there is an opportunity for the firm to interface with policymakers and regulators with the intention of influencing policy decisions or, on the other hand, the necessity to alter organizational processes and structures to accommodate existing or emerging regulations [7]. Mitigating these risks can help energy companies improve their financial position, stability and increase their robustness when operating in the complex and evolving business environment.

Table 3 presents a comprehensive overview of 15 cases from the energy sector, highlighting the impact of various financial indicators and risks on the companies' performance.

First, it has to be stated that the overall profitability differs from one company to another, some of them are growing, while the others face certain difficulties. This

Table 3 Comprehensive overview of 15 cases from the energy sector for the period from 2019 to 2023

№	Company	Period	Data	Description	Clarification
1.	Energy Co (USA)	2019	Profitability	New findings of Energy Co reveal that the net profit has risen to 15% more than the previous year	Net profit is one of the significant motivations of measuring the performance of business entities, as it defines the ability of a business to earn a profit
2.	Power Corp (Canada)	2020	Liquidity	Power Corp posted a lower current ratio of one, down from the prior year's value. 5 in 2020, which signal weakness in terms of liquidity in the business venture	A fragile liquidity ratio that aims to determine the capacity of a company to clear its short-term debts using its current assets. Any ratio less than 2 suggests that there might be some problem in the liquidity position of the firm
3.	Renewa Tech (Germany)	2021	Solvency	For the year ended 30 June, Renewa Tech's debt-to-equity ratio was raised to 1. Funded debt-to-total-capital ratio reflects the financial leverage of the company; it was 2 in 2021 which was comparatively higher	This one relates to the financial risk, and shows debt to equity—it indicates how much debt a company has in relation to equity. The reason is that when using a high ratio, the firm has more long-term debt as a source of financing than equity
4.	Power Gen (France)	2022	Market risk	Business sales of Power Gen reduced by 10% in 2022 because there were lower energy prices in the market	Market risk is the risk associated with the comparative fluctuation in the value of financial securities or qualities of the market

(continued)

Table 3 (continued)

№	Company	Period	Data	Description	Clarification
5.	Eco Energy (UK)	2019	Operational	Failure in the operations of a key facility also affected Eco Energy in the previous year through reduced production that was delayed for two weeks in 2019	Operational risks are risks to business that arise from various inadequate or failed activities that are conducted inside or outside the business entity
6.	Solar Solutions (China)	2020	Regulatory	Due to social issues in 2020, there was an issue of prolonged approval of the project that Solar Solutions offered	Legal risks are all those risks that are of legal nature such as the risks which arise as a result of change in laws and regulations affecting the company
7.	Wind Works (Japan)	2021	Market risk	Market demand for Wind Works in the market has been higher in the year 2021; hence, the firm reported higher sales and therefore a 20% rise in their revenue	Market risks can also be related to customer and competitor risks, which are the shifts in market environment
8.	Sun Power (India)	2022	Operational	Last year, Sun Power, a company in solar energy supply challenged by a supply chain breakdown thus causing a halt in production	Loosely, operational risks can be defined as resulting from commitments and operational activities or events that cause interruptions, delays or otherwise affect supply chain, accidents, or labor relations
9.	Bio Energy Co (Brazil)	2019	Regulatory	In 2019, the company was relying on the government subsidies for renewable energy projects, and it is based in Brazil	Other risks that can be within the scope of regulation risks are those that are linked to changes in policies with regard to taxes or subsidies or environmental legislations
10.	Gas Gen (Russia)	2020	Market risk	Therefore, we noted that Gas Gen was able to sell 5% less in 2020 attributed to the decreased use of natural gas in the market	Other related business risks encompass risks that relate with shifts in geographical risk or trade relations

(continued)

Table 3 (continued)

№	Company	Period	Data	Description	Clarification
11.	Hydro Power (Australia)	2021	Operational	Some of the business issues that affected Hydro Power in 2021 include; severe weather conditions affected the hydroelectric plants	Other operational risk factors that could be embraced range from natural calamities, weather conditions, or factors that are out of the organization's control among others
12.	Energy Solutions (South Korea)	2022	Market risk	A strategy that Energy Solutions applied in regards to its revenue in 2022 was to adjust its price in consideration to market competition	Market risks can also comprise risks that relate to the subject field of change in technology or the appearance of other market parties
13.	Thermal Energy (Italy)	2019	Regulatory	Environmental permits for Thermal Energy's thermal energy plants in Italy were a concern for it resulting in the delay	Such a view implies that regulatory risks may embrace risks related to legal cases or delays of projects' approval
14	Power Tech (Spain)	2020	Operational	There were other expenses in 2020 in connection with the unplanned maintenance shutdown of one of the company's power plants	Maintenance risks, for instance, may fall under operational risks together with other issues such as accidents or any other disruptions of operations
15.	Green Ener Co (South Africa)	2021	Regulatory	The external factors that affected Green Ener Co in the year 2021 include risks in changes of the renewable energy policies by the South African government	Other regulatory risks can also comprise of risks that relate to change in trade laws or international standards affecting the company's operations

Source Authors development using [22]

is an indicative of the different market environments and competitive structures in which these companies are operating. Second, there are liquidity-related risks in some cases, which proves that short-term liabilities should be well controlled to secure the company's sound financial status. Thirdly, solvency ratios reveal different degrees of financial risk, which affects the future performance and risk management plans.

External factors, which include but not limited to equipment breakdowns and supply chain problems, present a great threat to companies and affect their production capacity and financial performance. Another factor that cannot be overlooked is the regulatory risks which are the changes in the government policies and environmental laws and regulations that affect the strategies of firms and their financial performance. Other factors that define the sector include market risks on commodity prices and demand patterns, which make the sector ever-changing and dynamic, and thus call for companies to come up with strategic interventions.

From these cases, risk management and strategic management and learning from the best practices companies in the energy field can improve their preparedness for the challenges lying ahead in the energy field.

5.3 Financial Policies and Regulatory Environment

These financial policies are associated with energy carry finances and the policies impact the vitality of the operation of energy. Such measures are usually declared and implemented by the governments and other connected authorities with the aim to provide a correct regulation of the stimulation of investments and innovation activity on the one hand, and to maintain costs and stability of the financial systems on the other hand [15].

In the area of financial policy regulation, qualified energy price regulation is easily understandable. In most cases, governments come in to either set the price or to subsidize the price in efforts to ensure energy companies get to maintain their revenues high, while ensuring that cost incurred by the end user is still manageable. An unending stream of cash also means that the sector is able to support itself both in terms of funding as well as in its ability to attract more capital.

Regulatory organizations establish financial reporting standards, monitor regulatory compliance, and intervene to resolve market manipulation or fraud issues. Regulatory agencies help to maintain the sector's financial stability by providing openness and accountability.

Another point of focus in the field of the financial policy is to support renewable energy sources. Governments propose different bonuses as well as grants in relation to initiatives connected with renewable energy sources. Such policies go both for environmental efficiency and for the effective diversification of energy sources in order to renew the focus on fossil fuel and to increase the sector's balance in confrontation to market fluctuations [23].

Policies on the financial front that are put into practice contribute crucially in determining the achievement of viable financial position for the energy sector. These policies must be well formulated and then properly implemented by governments and other regulating authorities, so that on one hand it encourages investment and innovation while on the other it is not detrimental to the financial stability of the banking systems and other concerned bodies.

It is possible to compare the legal regulations in various countries and get valuable ideas about the work being done to maintain financial security in the energy industry.

The legal systems also differ from country to country, and this affects the dimensions of power generation firms and the amount of volatility they encounter.

One component of regulatory frameworks is the level of government involvement in energy price setting. The regulatory approach can have an impact on the sector's financial stability, with completely regulated markets potentially delivering more stability while providing less incentive for innovation and efficiency [7].

The other aspect is the one of renewable source of energy availability or lack of thereof. Such regulations may further endanger sector's financial sustainability by altering the competitiveness of renewable energy than to the conventional that is fossil-based energy.

Examine too, the types of regulatory frameworks/theories and how they differ in addressing of environmental regulations. On the same note, some nations have rigorous policies and standards of emission reduction and development of clean energy which may greatly affect the financial health of organizations that rely on the use of fossil fuel. Some have relatively loose policies meaning that the expenses for the energy companies may be on the low side though there could be some negative contributions to the environment.

The comparison of the specific country's legislation to those of other countries also presents a vivid example of the relation between the regulation and the solidity and sustainability of the energy sector. This way, best practices deriving from each country or state's experience can be implemented in other nations and policymakers and industry stakeholders can work together toward better regulation of long-term financial sustainability of the energy sector.

5.4 Strategies for Enhancing Financial Stability

Best practices for financial management in the energy sector encompass a range of strategies aimed at enhancing financial stability and sustainability. One key practice is effective risk management, including identifying and mitigating market risks, operational risks, and regulatory risks [1].

Controlling is also another important practice, especially the affair of financial planning and budgeting. Organizations should draw up sound financial strategies that can be applied under several scenarios and contingencies. It can help ensure that resources are used effectively and the organization is prepared for any emergent issues.

Innovation and technology improvements are critical to enhancing financial stability in the energy sector. For example, smart grid technology can increase energy distribution efficiency, while developments in renewable energy technologies can reduce dependency on fossil fuels and mitigate environmental dangers [13].

Innovation can also assist organizations in managing changes in the market and legal environment. The suggestions given in Table 4 highlight a roadmap for the policymakers, industry experts, and investors regarding the ways to promote stability and sustainability of the financial sector in the energy industry. These recommendations are derived from the analysis of the strengths and weaknesses of the sector and the best practices that are obtainable from the financial ratios, legal environment, and technology. Thus, following these recommendations, the stakeholders can assist in the advancement of the energy industry toward becoming more sustainable and coping with the future problems.

This provides policy makers and industries as well as the investment communities with a form of roadmap on how they want to manage this sector in a way that sees to it that sound policy on its financial plans and strategies are accomplished. By following such suggestions, the stakeholders can set out and progress the advancements in the diagnoses and paradigms of the energy economy. This is why it is very crucial to involve stakeholders and ensure their contribution in the attainment of these recommendations to result in the most effective and best utilization of the energy sector that will effectively promote contribution to the development of the economy as well as the preservation of the world's environment.

Table 4 Recommendations for policymakers, industry stakeholders, and investors

№	Stakeholder	Recommendation
1.	Policymakers	1.1. The legal frameworks need to be stable and predictable that will enable the creation of desired environment for investing in energy sector and at the same time encouraging innovations
		1.2. Support the use of renewable energy by putting incentives and subsidies to enable long-term renewable energy source usage to curb the usage of fossil-based energy
		1.3. Sustain financial stability in the energy sector via improvement in the regulation of the market and increasing accountability
2.	Industry associations	2.1. Work with relevant authorities to establish norms and modern practices relating to the financial administration of the sector's operations in the interest of boosting the overall financial health of the sector
		2.2. Encourage expenditure on research and development for the purpose of boosting inventions on the technology aspect, thus enhancing the operations efficiency and business financial performance
3.	Energy companies	3.1. Formulate detailed risk management plans for market risks, operation risks, and regulatory risks that threaten the financial stability
		3.2. Adopt modernism and spend on infrastructures that will enable the efficient use of energy and cutting on costs, hence being able to stand the competitive market
4.	Investors	4.1. Urges investors to take care on ESG factors while choosing energy firms as a way of investing for efficient and financially sound organizations
		4.2. Invest in renewable energy sources and technology that belong to the sustainable development goals and objective

Source Authors development

6 Conclusions

This research aimed at exploring the structure and development trends in the energy sector for 15 different countries of variable geographical location and economic diversification such as United States of America, Canada, Germany, France, United Kingdom, China, Japan, India, Brazil, Russia, Australia, South Korea, Italy, Spain, and Republic of South Africa. The analysis of the different factors affecting the sector's financial performance from 2019 to 2023 by the study uses ratio analysis, trends analysis, and an econometrics model on both qualitative and quantitative techniques.

The econometric model extracts influential factors that affect financial stability including EPV, RSI, GS, interest rate, GDP, and II. Fluctuations in energy price as analyzed point to less financial stability and hence have to come up with better risk management plans. From the results of the study, it can be concluded that stringent regulatory measures have a positive correlation with the financial position, thereby implying the need to continuously uphold standards of the global governance. Government subsidies and GDP growth notably assist financial sustainability implying the importance of economic policies in strengthening the sector's resistance.

The result of the regression analysis shows that the overall 69% of the model is accountable for the dependent variable. It indicates that 63% or (R-squared $= 0.6963$) of the total variances in the FSI for the 200 companies under analysis is explained by the independent variables. The key coefficients of the model have been achieved as EPV ($- 0.0297$), RSI (0.0524), GS (0.1373), GDP (0.2848), IR (0.1916), and II (0.0831). However, when energy price volatility increases in a country, then, it is implied to have negative effects on financial stability of the country as presented by the aforementioned coefficients. These coefficients imply that aspects such as increased regulatory stringency, provision of government subsidies, increased GDP, decline in interest rates, and technological advancement will enhance the financial stability of a country.

As reflected across all fifteen nations, the block's energy markets have attained different types of development. The nations that invested in the renewable power sector include China, USA, and Germany have relatively healthy and prospective sectors. Current high growth rate economies such as India and Brazil have recorded increased investment in the energy sector especially in renewable energy and policy support. Canada, France, and the UK show relatively conservative changes and steady growth due to the higher level of the developed economy and preferable legislation on the technologies' application.

The following are the recommendations that develop from this research study. It is therefore necessary to prevent the risk associated with volatile energy prices through effective risk management strategies. Techniques like hedging and diversification of energy sources may also improve the stability. Since this sector is rather sensitive to market fluctuations, constant introduction of new technologies and innovations is crucial in order to enhance the sector's stability. Policies should establish certainty for industry and strong financials; however, these aspects cannot be achieved without proper communication between policymakers and industry entities. Adhering to strict regulations is appropriate to foster order and check a number of stakeholders thereby enhancing stability of the financial systems.

The information that has been obtained in this research will help authorities to implement proper strategies with the aim to improve the quality of interventions that are related to financial stability. Another area that can be examined in further study is the effects of the transition to renewable energy sources on the stability of the energy sector and the influence of digitalization in the sector's potential future.

The present research emphasizes that multiple factors, which are of an economic, regulatory, and technological nature, define the future of the energy sector's financial sustainability. Thus, using quantitative and qualitative analyzes, the research provides answers to interesting questions that can be useful to stakeholders who want to function effectively in the modern environment for energy supply. Colleges and universities must balance overall risk management with focused policies and the ongoing development of technology and additional support in order to further promote the sector's future growth.

References

1. T. Kurbatova, D. Lysenko, G. Trypolska, O. Prokopenko, M. Järvis, T. Skibina, Solar energy for green university: estimation of economic, environmental and image benefits. Int. J. Glob. Environ. Issues 21(2–4), 198–216 (2022). https://doi.org/10.1504/IJGENVI.2022.126209
2. L.Ch. Muller, Hydrothermal liquefaction of spent coffee grounds followed by biocatalytic upgradation to produce biofuel: a circular economy approach. Biofuels (2021). https://doi.org/10.1080/17597269.2021.1948757
3. I. Bashynska, G. Smokvina, L. Yaremko, Y. Lemko, T. Ovcharenko, S. Zhang, Assessment of investment and innovation image of the regions of Ukraine in terms of sustainable transformations. Acta Innov. 43, 63–77 (2022). https://doi.org/10.32933/ActaInnovations.43.6
4. G. Trypolska, T. Kurbatova, O. Prokopenko, H. Howaniec, Y. Klapkiv, Wind and solar power plant end-of-life equipment: prospects for management in Ukraine. Energies 15(5), 1662 (2022). https://doi.org/10.3390/en15051662
5. J.V. Petlenko, B.P. Schehlyuk, Specific features of military technology marketing. Aktual'ni Problemy Ekonomiky Actual Probl. Econ. (160), 101 (2014)
6. D. Kretov, O. Mindova, B. Aitaliev, A. Koldovskyi, Development management of interbank competition in the corporate lending market. Econ. Ecol. Socium 7, 89–99 (2023).https://doi.org/10.31520/2616-7107/2023.7.2-7
7. M. Masyk, Z. Buryk, O. Radchenko, V. Saienko, Y. Dziurakh, Criteria for governance' institutional effectiveness and quality in the context of sustainable development tasks. Int. J. Qual. Res. 17(2), 501–514 (2023). https://doi.org/10.24874/IJQR17.02-13
8. K. Redko, O. Borychenko, A. Cherniavskyi, V. Saienko, S. Dudnikov, Comparative analysis of innovative development strategies of fuel and energy complex of Ukraine and the EU countries: international experience. Int. J. Energy Econ. Policy 13(2), 301–308 (2023). https://doi.org/10.32479/ijeep.14035
9. O. Prokopenko, T. Kurbatova, M. Khalilova, A. Zerkal, G. Prause, J. Binda, T. Berdiyorov, Y. Klapkiv, S. Sanetra-Półgrabi, I. Komarnitskyi, Impact of investments and R&D costs in renewable energy technologies on companies' profitability indicators: assessment and forecast. Energies 16(3), 1021 (2023)

10. O. Prokopenko, A. Chechel, A. Koldovskiy, M. Kldiashvili, Innovative models of green entrepreneurship: social impact on sustainable development of local economies. Econ. Ecol. Socium **8**, 89–111 (2024). https://doi.org/10.61954/2616-7107/2024.8.1-8

11. P. Bhola, Techno-economic and environmental assessment of utilizing campus building rooftops for solar PV power generation. Int. J. Green Energy (2021). https://doi.org/10.1080/15435075.2021.1904946

12. D. Gielen, The role of renewable energy in the global energy transformation. Curr. Opin. Environ. Sustain. **24**, 38–50 (2019)

13. S. Mukherjee, Harvesting energy from thunderstorm—a new source of renewable energy. Int. J. Adv. Sci. Eng. **6**(S1), 64–67 (2019)

14. L.Y. Dovhan, H.A. Mokhonyko, I.P. Malyk, *Project Management: A Textbook for Studying the Discipline for Masters of the Field of Knowledge 07 "Management and Administration" Specialty 073 "Management" Specialization: "Management and Business Administration", "Management of International Projects", "Management of Innovations", "Logistics"* (KPI named after Igor Sikorsky, Kyiv, 2017)

15. S.I. Pyrozhkov, V.V. Ryazantseva, R.M. Motorin et al., *Statistics: A Textbook* (Kyiv National University of Trade and Economics, Kyiv, 2020)

16. I. Sopronenkov, N. Zelisko, V. Vasylyna, I. Lutsenko, V. Saienko, Tax policy: impact on business development and economic dynamics of the country. Econ. Aff. **68**(04), 2025–2034 (2023). https://doi.org/10.46852/0424-2513.4.2023.14

17. M. Tymoshenko, V. Saienko, M. Serbov, M. Shashyna, O. Slavkova, The impact of industry 4.0 on modelling energy scenarios of the developing economies. Finan. Credit Act. Probl. Theory Pract. **1**(48), 336–350 (2023). https://doi.org/10.55643/fcaptp.1.48.2023.3941

18. A. Kuczabski, O. Aleinikova, H. Poberezhets, H. Tolchieva, V. Saienko, A. Skomorovskyi, The analysis of the effectiveness of regional development management. Int. J. Qual. Res. **17**(3), 695–706 (2023). https://doi.org/10.24874/IJQR17.03-05

19. I. Bashynska, L. Niekrasova, V. Osypov, A. Dyskina, L. Zakharchenko, Conceptual basis for the formation of a smart eco-industrial parks as benchmarking of sustainable manufacturing, in L. Moldovan, A. Gligor (eds.) *The 17th International Conference Interdisciplinarity in Engineering. Inter-ENG 2023. Lecture Notes in Networks and Systems*, vol. 928 (Springer, Cham, 2024), pp. 337–349

20. M. Dudek, I. Bashynska, S. Filyppova, S. Yermak, D. Cichoń, Methodology for assessment of inclusive social responsibility of the energy industry enterprises. J. Clean. Prod. **394**, 136317 (2023)

21. World Bank, World Bank Open Data (2024). https://data.worldbank.org

22. IMF, IMF Finance Data (2024). https://www.imf.org/en/Data

23. V. Mazur, A. Koldovskyi, L. Ryabushka, N. Yakubovska, The formation of a rational model of management of the construction company's capital structure. Finan. Credit Act. Probl. Theory Pract. **6**(53), 128–144 (2023). https://doi.org/10.55643/fcaptp.6.53.2023.4223

Compliance Management Implementation in Energy Sector Enterprises of the National Economy

Viktor Koval⬚, **Hanna Hrinchenko**⬚, **Anna Fomenko**⬚, **Natalia Didenko**⬚, and **Yana Medvedovska**⬚

Abstract The study proposes mechanisms for implementing compliance management in energy enterprises, considering relevant international and national security requirements. Approaches to sustainable energy development using renewable sources and proposed requirements for such energy equipment were analysed. To ensure the sustainable development of energy enterprises, the author proposes a compliance management model based on a structural and hierarchical system of regulatory support for safety and energy efficiency in the nuclear power sector, which forms a compliance policy. The mechanisms for implementing compliance management include an algorithm for compliance monitoring and control of the safety of power equipment operation, which is based on the proposed mathematical model for assessing technical conditions. The developed assessment models provide operational information on the actual current technical conditions and indicators of operational safety of power equipment, which allows constant monitoring of the technical condition of the equipment, regulation of the frequency of preventive maintenance and repairs to maintain the equipment in good condition, prediction of resources, and solving the issue of extending the service life or the expediency of equipment restoration, which in turn allows effective management of the energy enterprise and ensures equipment safety.

H. Hrinchenko (✉)
Educational and Research Institute 'Ukrainian Engineering-Pedagogics Academy', V. N. Karazin Kharkiv National University, 16 Universiteyskaya St., Kharkiv 61013, Ukraine
e-mail: hrinchenko@uipa.edu.ua

V. Koval
Izmail State University of Humanities, 12 Repina Str., Izmail 68610, Ukraine

A. Fomenko
Wismar University of Applied Sciences Technology, Business and Design, 14 Philipp-Müller-Straße, Wismar 23966, Germany

N. Didenko · Y. Medvedovska
Kharkiv National Automobile and Highway University, 25 Yaroslava Mudrogo Str., Kharkiv 61002, Ukraine

Keywords Compliance management · Sustainable development · Renewable energy · Safety assessment · Power equipment

1 Introduction

Ensuring sustainable development of enterprises and organizations is possible only if modern management solutions based on a systematic approach that covers both technical and social aspects are implemented. Compliance management is one of the most effective management approaches that involves efficient organization of processes and a system of monitoring (control) of operational risks. A key component of compliance management is ensuring compliance with laws and regulations, which includes the organization's regulatory framework, as well as national and international regulations that affect various aspects of business processes [1]. An organization's comprehensive compliance management programme can effectively manage compliance risks, adhere to ethical and environmental standards, and maintain the trust of stakeholders, including customers, regulators, investors, and the public, while focusing on achieving the Sustainable Development Goals (SDGs).

In this context, nuclear power, with the implementation of the management mechanisms, from the linear to the circular (renewable) model, plays a crucial role in providing a stable and powerful source of energy, making it possible to gain energy independence, increase economic sustainability, and promote environmentally friendly growth. As one of the most critical sectors in Ukraine's national economy, nuclear energy requires implementing effective management solutions for enterprises in this sector. The operation of equipment at nuclear power plants involves increased safety requirements, as the specificity of electricity generation at nuclear power plants (NPPs) is the use of nuclear fuel, which makes them particularly dangerous. The NPP operation experience shows that making wrong management decisions can cause significant economic and environmental losses and threaten human life. This issue is especially acute when reassigning the service life, which is a normal practice for nuclear power plants around the world. The design lifetime of an NPP power unit is 30 years, after which the question of further operation arises: the extension of the lifetime or decommissioning of the power unit [2, 3]. NPP lifetime extension is associated with increased requirements for reliability and safety of equipment operation, taking into account environmental aspects and energy efficiency, which should be reflected in regulatory and guidance documents. To ensure the safe operation of NPP equipment beyond the design life, it is necessary to develop a system of regulatory documents that reflects the requirements for effective management of a power facility, taking into account changes in operation under the influence of external and internal factors. Economic losses from incorrect decisions on the decommissioning of a particular NPP unit or on an unreasonable extension of the designated service life of its equipment are significant and affect the entire economy of the country. Developing an effective compliance management system, including a compliance policy, regulatory support for assessing the technical condition of equipment and the safety of operating processes, assessing the risks of the operation and ensuring compliance policy, justifying the possibilities of operating energy equipment, and taking into account energy efficiency and social responsibility, can become an effective mechanism for managing an energy company of any size. The implementation of compliance management will make it possible to identify and respond to gaps and critical points in the existing management and regulatory system.

2 Literature Review

One of the priorities of the national economy of any country is to ensure efficient management of energy resources, which takes into account social, economic, and environmental factors and enables sustainable development. Management decisions should have a systematic approach and consider social responsibility at all levels and processes, so it is extremely important to actively implement the concept of "compliance" in everyday life.

Implementation of compliance management by top management is highly related to organizational culture and compliance goals [4–6]. This study presents an operational approach to defining goals as well as creating and monitoring programmes aimed at achieving them. Enterprise risk management (ERM) practices are particularly important for ensuring stable regulatory compliance and have a positive and significant impact on both financial and operational efficiencies. An IT strategy also has a beneficial impact on a company's productivity.

The three pillars of intellectual capital (human, structural, and relational) are related to enterprise innovation and how successful knowledge management can improve business productivity, innovation, and environmental standards and are components of compliance management [7–9]. In addition, corporate innovations and successful knowledge management [10], namely training personnel from administration to practitioners (production workers) in social responsibility and compliance with environmental standards, significantly increase business efficiency.

The impact of corporate social responsibility on environmental changes was studied in researches [11–13] based on the data of large and medium-sized enterprises. The results presented by the authors show that corporate social responsibility does not have a direct impact on environmental parameters, but positively correlates with the strategy of environmental development and green innovations, which in turn are aimed at improving environmental parameters. That is, it is a significant mediator between them. This study proposes a model for implementing compliance management elements for CEOs of manufacturing organizations and policy-makers to manage corporate social responsibility, environmental strategy, and green innovations in environmental performance analysis.

The strategy for implementing effective management in energy enterprises is based on management approaches using geoinformation and the environmental aspects of operation [14–16]. The combination of effective economic management and sustainability of an enterprise in the context of a shift to a circular economy in energy and renewable production was considered [17–19]. The proposed models, approaches, strategies, and plans are based on ensuring the economic sustainability of energy facilities. At the same time, it is emphasized that implementing such mechanisms is impossible without considering the factors affecting the safety of the operation.

In subsequent studies, assessments of the technical condition and extension of the service life of energy equipment were used to improve the management of a power facility and ensure energy efficiency, as well as the safety of individual equipment (pipeline systems, power unit structures, electromechanical equipment, etc.) [20–25]. A discussion of certain technical diagnostics of power equipment is presented, which should become part of comprehensive regulatory support for the safety of power equipment operation. Research methods based on calculations using a mathematical model for assessing the actual state of power equipment and comparing the calculation results are proposed, which allows the prediction of the limit parameters of the technical condition and monitoring of mechanical properties, as well as determining the residual life of structures, considering geotechnical and seismotectonic conditions [23–25].

Several indicators have been proposed to determine the safety and quality of power equipment operations, which are normative indicators for relative assessment. Based on the proposed approaches for assessing the quality of multi-parameter energy facilities, normative criteria for the optimality of technical and economic indicators are proposed. At the same time, the authors propose defining a criterion for assessing the quality indicators of the overall system of low-potential NPP complexes and specific objects, such as reactor fuel elements.

Another aspect of ensuring the safe operation of energy facilities is diagnosing and assessing risks, which is vital to consider and is an integral part of energy compliance management [26–35]. The authors proposed models for risk assessment at nuclear power facilities, taking into account production and social factors and models for forecasting the impact of these factors on the environment and economic processes.

To summarize, it can be said that certain aspects of compliance management are covered in the works of scholars. At the same time, there is no systematization and no unified concept of mechanisms for implementing compliance management at energy facilities. Compliance management of an energy company requires considering regulatory support at various levels, technical aspects of operation with regard to safety and energy efficiency, impact on the environmental component and social responsibility, calculation of risks of production, and implementation of the compliance approach, and development of personnel training programmes on the principles of compliance management (i.e. management of knowledge and awareness of employees).

3 Research Methodology

Nowadays, Ukraine ranks seventh in the world regarding nuclear power generation and third in the nuclear power share in the country's total energy balance. Nuclear power plants account for 55% of total electricity generation in Ukraine. Compared to other countries, more than 25% of energy balances in thirteen countries depend on nuclear energy. In France, nuclear power accounts for more than 70% of the energy balance, Slovakia for more than 53%, and Hungary and Bulgaria for more than 40%.

The development of an effective management strategy for an energy company, like any other structure, should be based on the national economic development strategy and international policy, which are supported by international and European regulatory documents and by which internal standards, regulations, and practices have been developed. It is possible to consider different levels of regulatory support in the activities by implementing compliance management approaches. This position should be supported by continuous monitoring and improvement (Fig. 1), which is a requirement of the ISO 37000: 2021 standard [36].

Implementation of compliance management approaches in an energy company can be an effective way and have the following implementation mechanisms: creation of compliance policies and procedures under legislation, standards, and ethical norms; organization of a control and audit system; creation of internal control bodies; education and training of personnel on compliance, their rights, and obligations; as well as procedures and standards to be followed; ensuring access to information on compliance requirements and procedures for their implementation for all stakeholders. The compliance management system is aimed at continuous improvement based on risk assessment, internal control, and audits. In the compliance management system of energy companies, risk assessment and safety of power equipment are incorporated at every stage and are the basis for making any technical and managerial decisions. Control and monitoring compliance with regulatory safety requirements is included in the "control mechanisms" stage and is a crucial system element.

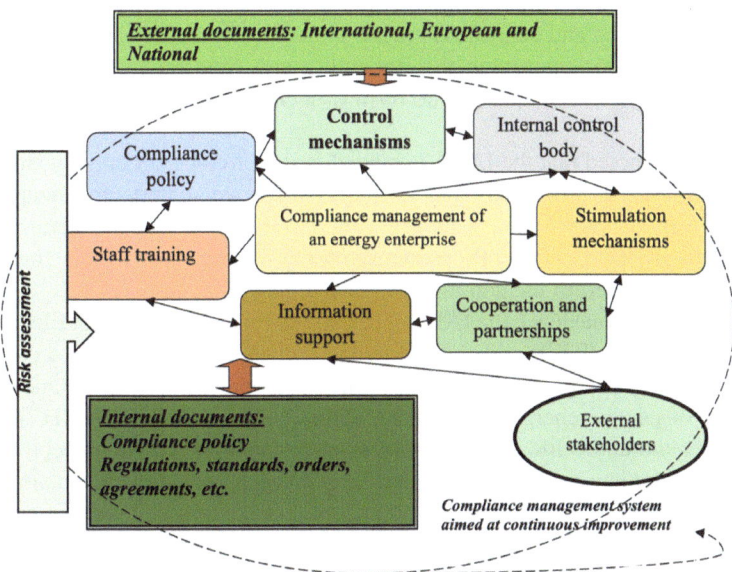

Fig. 1 Model of compliance management of an energy company

Safety is maintained by assessing the operational condition of power equipment according to Ukrainian regulations and recommendations of the International Atomic Energy Regulatory Organization (IAEA). IAEA safety standards are divided into several areas:

- Specific Safety Guides (SSGs): Documents that provide specific guidance and recommendations on certain safety aspects in the nuclear industry.
- General Safety Guides (GSGs): Documents that provide general safety principles and guidance that apply broadly to various aspects of nuclear activities.
- Specific Safety Requirements (SSRs): Documents that specify specific safety requirements and standards that must be met in specific situations or for specific types of nuclear facilities.
- General Safety Requirements (GSRs): Documents that specify general safety principles and requirements generally applicable to various aspects of nuclear safety.
- Safety Fundamentals (SFs): Documents that establish the fundamental principles and safety basis on which the entire system of IAEA standards and guidance on nuclear safety is based.

These different areas of IAEA standards interact to create a comprehensive system of regulations to ensure safety in the nuclear industry. In addition, standards for managing the ageing of power equipment and extending its service life can be distinguished, such as [37–40]. At the same time, there are no separate standards that ensure the efficiency of equipment operation, significantly when the service life is extended or during the off-design life of equipment, and there are no regulatory indicators in the IAEA system. However, these requirements are very important for effective monitoring and control. There are some developments, such as "Advanced Control Systems for Improving the Reliability and Efficiency of Nuclear Power Plants", which was developed in the form of technical documentation (TECDOC Series) [41].

It should be noted that the IAEA is guided in its activities not only by standards. The regulatory framework of the International Atomic Regulatory Organization also includes technical documents, technical reports (Technical Reports Series), panel discussions and reports (Panel Proceedings Series). The entire regulatory framework is advisory but is part of the compliance management system.

At the international level, the IAEA cooperates with the International Electrotechnical Commission (IEC), and their activities are aimed at the safe, secure, and peaceful use of nuclear technologies and the development of global standards for the safety of nuclear energy (Technical Committee, TC 45). The IEC TC 45 standards cover the entire life cycle of nuclear power systems, from concept to engineering, production, testing, installation, commissioning, operating, maintaining, ageing management, modernization, and decommissioning. IEC's core business is instrumentation, control systems, and power systems, which are essential for the safety of nuclear power plants. The nuclear sector has its own well-developed safety policy and philosophy. Hence, IEC safety publications consider differences from the general approach and provide guidelines specific to nuclear-related facilities with an integrated approach to safety.

In accordance with the TC 45/IAEA agreement, the IEC Nuclear Sector Safety and Security Standards implement concepts and terminology of the IAEA Safety and Security Guides. The main scope of activities includes instrumentation used for monitoring, control, and safety management functions.

The IEC develops both general standards (horizontal) [42] and specific standards (vertical) [43]. Horizontal standards, also known as core standards, are technology-independent. They are applicable in many technical fields. Vertical standards are intended to meet specific technical requirements.

The working groups of Technical Committee TC45 have two subcommittees, which in turn are divided into working groups (WG), whose activities are aimed at a certain aspect of standardization in the field of nuclear energy: sensors and measurement methods; measuring instruments and control systems: architecture and system aspects; measurement of processes and radiation monitoring; functional and safety basics of measuring, control and power systems; control rooms, human–machine interfaces and human factor engineering; performance and resilience of systems to external influences; ageing management of instrumentation, control and power systems at NPPs; electric power systems: architecture and system aspects; artificial intelligence for nuclear facilities.

For example, WG7 Functional and Safety Fundamentals for Instrumentation, Control and Power Systems (WG7) develops and maintains standards and reports on selected aspects of fundamental principles for instrumentation, control and power systems relevant to the safety of nuclear installations (i.e. power plants, nuclear fuel cycle and waste management facilities).

The standards developed by WG7 extend the IAEA standards and guidelines, including the IAEA's Safety Framework:

- SF-1 Fundamental Safety Principles [44].
- SSR-2/1 Safety of Nuclear Power Plants-Design [45].
- NS-R-4 Safety of Research Reactors [46].
- NS-R-5 (Rev. 1) Safety of Nuclear Fuel Cycle Facilities [47].
- INSAG-10 "Advanced Protection in the Field of Nuclear Safety" [48].

Thus, the task of WG7 includes:

- Categorization of safety functions.
- General principles for the classification of safety-relevant instrumentation, control systems, and power systems and requirements for separation of functions.
- Depth protection.
- Denial of service in case of single failures.

Protection against common cause failures (including systematic failures) and implementation of methods such as separation, diversification, and isolation to achieve independence and protection against internal and external hazards, as well as random and systematic failures.

- Quantification of reliability, its assurance through observation and verification tests, and its limitation by taking into account failures due to common causes and systematic failures.
- Analysis methods for identifying and eliminating potential random and systematic failures.

One of the main principles of regulatory and legal regulation in Ukraine is a systemic and hierarchical approach to developing and updating regulatory documents for nuclear power plants (NPPs). In practice, this principle is implemented by creating a hierarchical structure of regulatory documents, including documents of several levels, presented in Fig. 2 and is the basis for the compliance policy of energy sector enterprises.

The structure includes three standards in force in Ukraine: international, European and national (Fig. 2), and the leading organizations that develop standards at these levels. The organizations are conditionally divided into those developing industry-specific standards (in the nuclear power industry) and general and industry-specific standards. "General" standards (horizontal) are understood as regulatory documents applied not only in the nuclear power industry but can be used in other areas of activity. Both sectoral and general national standards are developed based on inter-national and European standards and recommendations. For systematization, were selected regulatory documents that apply to NPPs and nuclear facilities and are aimed

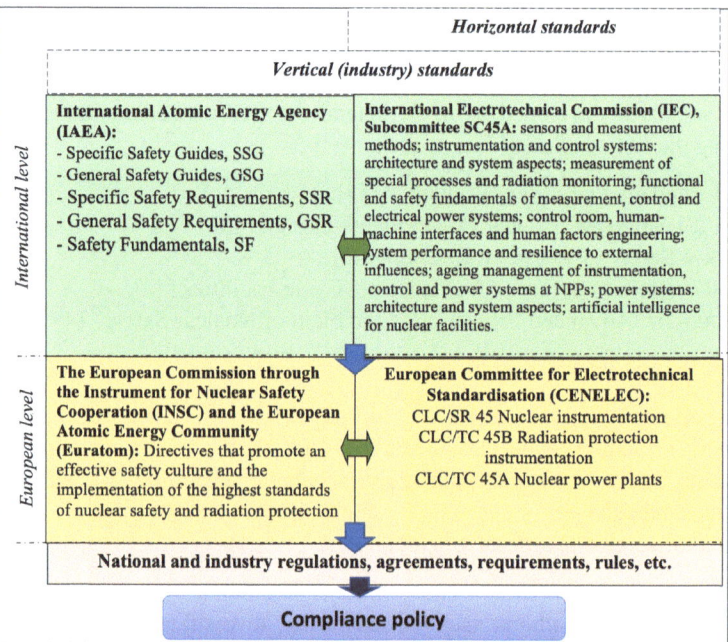

Fig. 2 Structural and hierarchical system of regulatory support for safety and efficiency in the nuclear energy sector, which forms the compliance policy

at ensuring nuclear and radiation safety of NPPs in general, standards, norms and rules of technological processes, unified control methods, including industry standards, standard programmes according to which maintenance, testing, assessment of technical condition and forecasting of equipment residual life are performed; regulatory documents that define requirements for analysis of operational reliability and energy efficiency of facilities. Regulatory documents of all levels form the basis of the energy company's compliance policy, which is based on the principle of compliance with the highest standards of nuclear and radiation safety to ensure safe electricity generation.

Analysing the existing regulatory framework for energy companies' compliance policies, the study reveals a number of issues that need to be improved to enhance management and security. In the above standards, the condition of the equipment is qualitatively assessed, for example, as "serviceable" or "faulty", "operable" or "inoperable". This approach is unacceptable for the concept based on the "safe operation by technical condition" principle. The traditional methods of monitoring and assessing the quality of equipment recommended by the standards solve the problems of current operation without considering the accumulation of degradation changes associated with the ageing processes of equipment during operation. As a result, it is challenging to apply ineffective management when monitoring and controlling within the compliance management framework.

4 Results and Discussion

In order to assess the safety of functioning of an energy enterprise, it is proposed to improve the compliance management system with regulatory approaches to assessing the technical condition by the parameters "safety", "residual life", "forecasting the residual life", and, accordingly, to carry out risk management of the enterprise according to these parameters. Based on diagnostics of the technical condition of equipment with the possibility of predicting the residual life of equipment, the author proposes model (1), the essence of which is to determine the general form of the regression equation of technical parameters of the equipment, to build estimates of the parameters included in its equation, and to test statistical hypotheses about the regression of parameters over time. This model makes a forecast by extrapolating the values of the dominant parameter based on statistical data for the period from the operation's start to the study's time. According to the regression theory, let's assume that the dependence of a random variable Y (the dominant determinant of equipment quality or its K_D, calculated from statistical data for the past time period) on another property variable X (operating conditions) is registered on a set of points x_n by a set of values y_n, while the values x_n and y_n reflect the actual values of $Y(x_n)$ with a random error Z_n. Then, the regression of y on x is as follows:

$$Y_n = f(x_n, a_0, a_1, \ldots, a_k) + Z_n, \tag{1}$$

where a_0, a_1, ..., a_k denotes the set of parameters that determine the function $f(x)$. The error Z_n is estimated by the least-squares method, which, assuming a normal distribution of the observations, leads to estimates for Z_n that coincide with the estimates of the highest probability.

As a result, a model is adopted that minimizes the sum of the squares of deviations calculated using the formulas for approximating the values of the K_D coefficient during the periods of operation between scheduled maintenance, during which the parameters are measured.

Based on these parameters with the dominant coefficients of relative assessment of the technical condition K_D, approximating models of their change are built. The extrapolation function extends the "past" values of the parameter into the "future" (Fig. 3). For the time-series model, time is the independent variable. The statistical data of the parameter are plotted on the graph according to time. Then, a curve is selected using this least-squares data, which continues into the "future".

The use of extrapolation makes it possible to obtain a picture of the further change in the K_D coefficient over time and obtain the value of the residual life by the graph-analytical method.

The predicted value of the residual service life of the equipment is determined under the assumption that the trend of the ageing intensity at the end of the design life will be maintained during the post-design life. At the same time, the average rate of change of the defining parameter or coefficient K_D for the last period before the inspection will be maintained during further operation. This assumption is approximate but is confirmed by investigations on the equipment in operation.

Taking into account the above, the estimated value of the remaining lifetime is as follows:

$$T_{rem} = \frac{t}{\Delta K_D}(K_D \text{ meas} - 0.2), \tag{2}$$

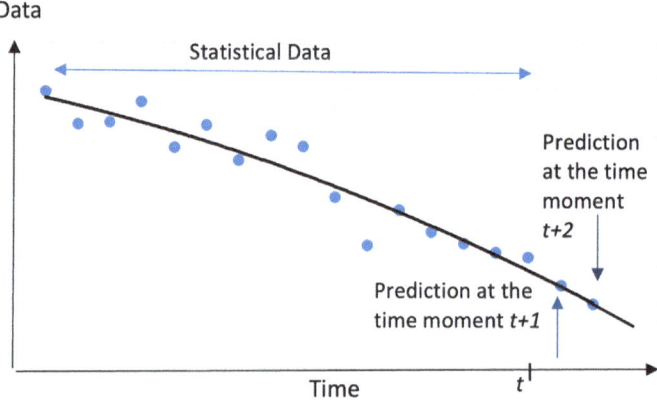

Fig. 3 Approximation model of the change in the dominant coefficients

where

t is the time from the previous measurement to the current one.

$|\Delta K_D|$ is the change in the value of the dominant coefficient over time t.

K_D meas is the dominant coefficient of the technical condition during the measurement period, which is determined using a multi-parameter mathematical programming model.

0.2   is the technical condition coefficient determined experimentally in the calculations as the limit under the conditions of storage of the required safety equipment performance margin in the post-project operation period.

The use of the graph-analytical method for predicting the residual lifetime based on a preliminary assessment of the technical condition, taking into account the physical nature of the predicted processes, allows us to draw scientifically based decisions on the value of the final lifetime of equipment in each case separately.

The average time to failure (T_{av}) is determined by the formula:

$$T_{av} = \frac{(P - P_0)}{a} \cdot \left(1 + \frac{v^2}{2}\right), \tag{3}$$

where

a is the average rate of change of the dominant parameter.

P_0 is the initial value of the dominant parameter.

P is the limit value of the dominant parameter.

$|v|$ is the coefficient of variation of the main degradation processes, which for the equipment with the lowest value of the K_D coefficient at the time of the investigation is determined by the formula:

$$v = \frac{\sqrt{n}}{\sum_{i-1}^{n} K_D} \sqrt{\sum_{i=1}^{n}\left(K_D - \frac{1}{n}\sum_{i=1}^{n} K_D\right)^2}, \tag{4}$$

where

n is the number of samples of the parameter for the previous period.

The gamma percentage operating rate ($T\gamma$) is determined by the formula:

$$T_\gamma = \frac{(P - P_0)}{a}\left(1 + \frac{v^2 u_\gamma^2}{2} - v \cdot u_\gamma \sqrt{1 + \frac{v^2 u_\gamma^2}{4}}\right), \tag{5}$$

where

u is the quantile of the normative normal distribution at the regulatory (specified) failure probability $\gamma\% = 95\%$.

The probability of failure-free operation $P(t)$ is determined for a monotonic function by a simplified formula:

$$P(t) = \Phi\left[\frac{(P - P_0) - at}{v \cdot \sqrt{at(P - P_0)}}\right] \tag{6}$$

In the DN-range, when the degradation process is non-monotonic, the mean time between failures (T_{av}) is determined by the formula:

$$T_{av} = \frac{(P - P_0)}{a} \tag{7}$$

The gamma percentage operating rate $(T\gamma)$ is determined:

$$T_\gamma = \frac{(P - P_0)}{a} \cdot \chi\left(1 - \frac{\gamma}{100}v\right). \tag{8}$$

The probability of failure-free operation $P(t)$ is determined for a non-monotonic function by the formula:

$$P(t) = \Phi\left[\frac{(P - P_0) - at}{v \cdot \sqrt{at(P - P_0)}}\right] - e^{\frac{2}{v^2}} \cdot \Phi\left[-\frac{(P - P_0) + at}{v \cdot \sqrt{at(P - P_0)}}\right] \tag{9}$$

An algorithm for assessing the safety of power equipment operation has been developed to implement the proposed methods for compliance monitoring (Fig. 4).

For example, Table 1 presents statistical data on changes in the coolant temperature in the main circulation pump for 2018–2023 at the South Ukrainian NPP (Ukraine) and its assessment by the dominant indicator. As can be seen, in 2022, the temperature increase to 295 °C reduced the technical condition of the equipment to "satisfactory" $(K_D = 0.49)$. Adjusting this parameter made it possible to restore the technical condition to "good".

The developed methods for assessing the technical condition of power equipment make it possible to obtain at any time information from existing NPP information systems on the actual current technical condition and operational reliability of the leading electrical equipment of a power unit, which allows obtaining continuous monitoring of the technical condition of the equipment; regulate the frequency of preventive maintenance and repairs to maintain the equipment in good condition; plan the stock; resolve the issue of extending the service life or expediency of equipment replacement, which in turn allows to effectively manage the energy company, as well as ensure the safety of its operation within the framework of compliance management.

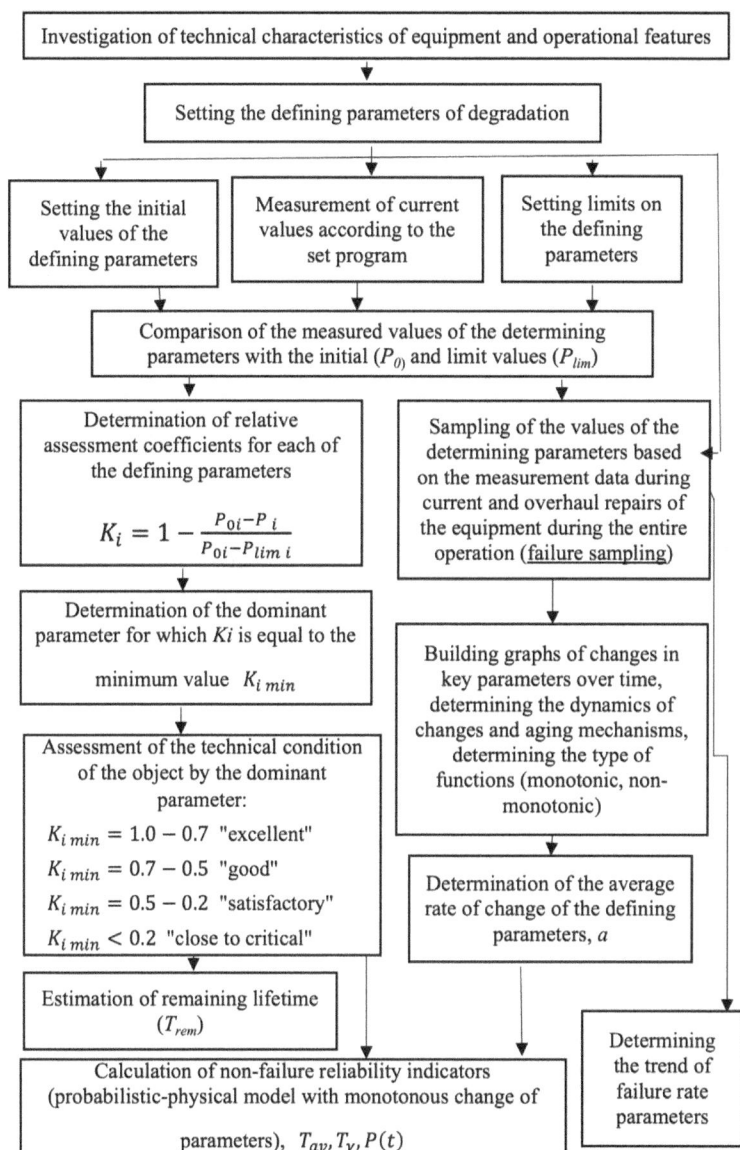

Fig. 4 Algorithm for compliance monitoring and control of the safety of power equipment operation

Table 1 Results of the technical condition assessment by the dominant indicator

Parameter	Year of measurement					
	2018	2019	2020	2021	2022	2023
Coolant temperature, °C	280	287	289	282	295	288
Dominant coefficient, K_D	0.68	0.6	0.58	0.65	0.49	0.59
Qualitative assessment of the technical condition by the dominant indicator	Good	Good	Good	Good	Satisfactory	Good

5 Conclusion

Developing an effective strategy for implementing compliance management in an energy company is only possible by considering regulatory support for the functioning of both general and technical processes. Having analysed the existing regulatory support of energy companies, such as nuclear power facilities, it can be concluded that the existing regulatory, organizational, methodological, and scientific approaches are sufficiently covered in standards and regulatory documents at various levels. These regulatory documents have become the basis for an energy company's compliance policy and, simultaneously, need to be improved to implement genuinely effective compliance management. After analysing the primary regulatory documents defining the criteria and principles of safe operation of nuclear power plant equipment, it can be noted that the basis of monitoring and control in the compliance management system is the need to constantly control the equipment's technical condition to ensure reliable operation and timely decision-making on further operation of elements and structures or their replacement.

After reviewing and analysing the regulatory documentation on ensuring the safe operation of NPPs, a number of shortcomings were identified, and, in this regard, improvements were proposed, taking into account the existing criteria and requirements for safe operation imposed on the power equipment of nuclear power plants. To this end, a model for assessing the technical condition is proposed, which forms the basis of the compliance monitoring and control algorithm and establishes the procedure, methods, means, and methods for studying power equipment, conducting a calculated assessment of the condition under long-term operation, and determining the remaining resources.

The procedure and methodology for assessing the technical condition of equipment under the influence of various operational factors, the procedure for obtaining information to assess the technical condition of the equipment under investigation, determining the residual life to decide on the possibility of further safe operation, etc., will form the basis for the development of regulatory documentation for compliance monitoring and control. The object of resource assessment and forecasting can be any energy facility at different stages of the operational cycle.

Further research will involve creating a compliance monitoring and control database that will include the following information: equipment operating data, technical performance measurement data, results of calculated performance analysis, data on technical measures, and management decisions. The development of implementation mechanisms will include developing programmes for the design of programmes and training personnel in compliance management principles.

References

1. W. Li, Y.A. Zhang, X. Li, Reducing symbolic compliance: the presence of multiple large shareholders as an internal monitoring mechanism. J. Manag. Stud. **61**(5), 1946–1984 (2024). https://doi.org/10.1111/joms.12961
2. International Atomic Energy Agency, Regulatory oversight of ageing management and long term operation programme of nuclear power plants, in *Safety Reports Series No. 109* (IAEA, Vienna, 2022)
3. International Atomic Energy Agency, Plant life management models for long term operation of nuclear power plants, in *IAEA Nuclear Energy Series No. NP-T-3.18* (IAEA, Vienna, 2015)
4. B. Zoellick, T. Frank, Governance, risk management, and compliance: an operational approach, in *Compliance Consortium Whitepaper* (2005) [online]. Available at http://www.securitymana gement.com/archive/library/compliance_consortium0805.pdf
5. H.M. Wee, M.F. Blos, W.-H. Yang, Risk management in logistics, in *Handbook on Decision Making* (Springer, Berlin, Heidelberg, 2012), pp. 285–305
6. M.L.D. Tewu, I. Bernarto, S. Suwarno, P. Lisdiono, The effect of enterprise risk management and compliance practices on the firm performance of Indonesian banking companies. Indonesian J. Bus. Entrepreneursh. **10**(1), 52 (2024). https://doi.org/10.17358/ijbe.10.1.52
7. B.T.T. Truong, P.V. Nguyen, D. Vrontis, Z.U. Ahmed, Unleashing corporate potential: the interplay of intellectual capital, knowledge management, and environmental compliance in enhancing innovation and performance. J. Knowl. Manage. **28**(4), 1054–1073 (2024). https://doi.org/10.1108/JKM-05-2023-0389
8. B.T.T. Truong, P.V. Nguyen, D. Vrontis, I. Inuwa, Exploring the interplay of intellectual capital, environmental compliance, innovation and social media usage in enhancing business performance in Vietnamese manufacturers. J. Intell. Capital (2024). https://doi.org/10.1108/JIC-10-2023-0233
9. S.S. Ahmed, J. Guozhu, S. Mubarik, M. Khan, E. Khan, Intellectual capital and business performance: the role of dimensions of absorptive capacity. J. Intell. Capital **21**(1), 23–39 (2019). https://doi.org/10.1108/jic-11-2018-0199
10. J. Alegre, K. Sengupta, R. Lapiedra, Knowledge management and innovation performance in a high-tech SMEs industry. Int. Small Bus. J. **31**(4), 454–470 (2013). https://doi.org/10.1177/0266242611417472
11. S. Kraus, S.U. Rehman, F.J.S. García, Corporate social responsibility and environmental performance: the mediating role of environmental strategy and green innovation. Technol. Forecast. Social Change **160**(2020), 120262 (2020). https://doi.org/10.1016/j.techfore.2020.120262
12. N. Anwar, N.H. Nik Mahmood, M.Y. Yusliza, T. Ramayah, J. Noor Faezah, W. Khalid, Green human resource management for organisational citizenship behaviour towards the environment and environmental performance on a university campus. J. Clean. Prod. **256**, 120401 (2020). https://doi.org/10.1016/j.jclepro.2020.120401
13. T.J. Arrive, M. Feng, Y. Yan, S.M. Chege, The involvement of telecommunication industry in the road to corporate sustainability and corporate social responsibility commitment. Corporate Soc. Responsib. Environ. Manage. **26**(1), 152–158 (2019). https://doi.org/.1002/csr.1667
14. M. Geissdoerfer, P. Savaget, N. Bocken, E.J. Hultink, The circular economy—a new sustainability paradigm? J. Clean. Prod. **143**, 757–768 (2017)

15. O. Ostapenko, P. Olczak, V. Koval, L. Hren, D. Matuszewska, O. Postupna, Application of geoinformation systems for assessment of effective integration of renewable energy technologies in the energy sector of Ukraine. Appl. Sci. **12**, 592 (2022). https://doi.org/10.3390/app120 20592

16. M.S.S. Danish, P. Gábor, Environmental and economic efficiency of nuclear projects. Clean Energy Invest. Zero Emission Projects, 115–126 (2022). https://doi.org/10.1007/978-3-031-12958-2_10

17. V. Koval, V. Khaustova, S. Lippolis, O. Ilyash, T. Salashenko, P. Olczak, Fundamental shifts in the EU's electric power sector development: LMDI decomposition analysis. Energies **16**(14), 5478 (2023). https://doi.org/10.3390/en16145478

18. V. Koval, I.W.E. Arsawan, N.P.S. Suryantini, S. Kovbasenko, N. Fisunenko, T. Aloshyna, Circular economy and sustainability-oriented innovation: conceptual framework and energy future avenue. Energies **16**, 243 (2022). https://doi.org/10.3390/en16010243

19. H. Hrinchenko, O. Prokopenko, N. Shmygol, V. Koval, L. Filipishyna, S. Palii, L.-I. Cioca, Sustainable energy safety management utilizing an industry-relative assessment of enterprise equipment technical condition. Sustainability **16**, 771 (2024). https://doi.org/10.3390/su1602 0771

20. H. Hrinchenko, O. Kupriyanov, V. Khomenko, S. Khomenko, V. Kniazieva, An approach to ensure operational safety for renewable energy equipment, in *Circular Economy for Renewable Energy. Green Energy and Technology*, eds. by V. Koval, P. Olczak (Springer, Cham, 2023), pp. 1–17. https://doi.org/10.1007/978-3-031-30800-0_1

21. H. Hrinchenko, V. Koval, N. Shmygol, O. Sydorov, O. Tsimoshynska, D. Matuszewska, Approaches to sustainable energy management in ensuring safety of power equipment operation. Energies **16**(18), 6488 (2023). https://doi.org/10.3390/en16186488

22. H. Hrinchenko, R. Trisch, V. Burdeina, S. Chelysheva, Algorithm of technical diagnostics of the complicated damage to the continued resource of the circulation pipeline of the nuclear power plant. Probl. Atom. Sci. Technol. **2**(120), 104–110 (2019)

23. T. Liu, Z. Wu, M. Bensi, Z. Ma, A mechanistic model of a PWR-based nuclear power plant in response to external hazard-induced station blackout accidents. Front. Energy Res.**11** (2023). https://doi.org/10.3389/fenrg.2023.1191467

24. J. Králik, Actual problems of the safety and reliability of the NPP structures in Slovakia. Key Eng. Mater. **738**, 261–272 (2017). https://doi.org/10.4028/www.scientific.net/KEM.738.261

25. A. Elbayoumi, T. Tahvonen, Novel methodology for functional design chain analysis of a nuclear power plant: a new built Finnish power plant case study. Nucl. Eng. Des.**393** (2022). https://doi.org/10.1016/j.nucengdes.2022.111795

26. Z. Hózer, M. Adorni, A. Arkoma, V. Busser, B. Bürger, K. Dieschbourg, R. Farkas, N. Girault, L.E. Herranz, R. Iglesias, M. Jobst, A. Kecek, C. Leclere, R. Lishchuk, M. Massone, N. Müllner, S. Sholomitsky, E. Slonszki, P. Szabó, T. Taurines, R. Zimmerl, Review of experimental database to support nuclear power plant safety analyses in SGTR and LOCA domains. Ann. Nucl. Energy **193**, 110001 (2023). https://doi.org/10.1016/j.anucene.2023.110001

27. H. Hrinchenko, O. Kupriyanov, R. Trishch, N. Antonenko, T. Bubela, Assessment of the quality of operation of equipment of nuclear power plants for the purpose of safe green transformation. AIP Conf. Proc. **3051**(1), 100004 (2024). https://doi.org/10.1063/5.0191649

28. R.O. Meyer, W. Wiesenack, A critique of fuel behavior in LOCA safety analyses and a proposed alternative. Nucl. Eng. Des. **394**, 111816 (2022). https://doi.org/10.1016/j.nucengdes.2022.111816

29. S. Ciattaglia, G. Federici, L. Barucca, R. Stieglitz, N. Taylor, EU DEMO safety and balance of plant design and operating requirements. Issues and possible solutions. Fusion Eng. Des.**146**(Part B), 2184–2188 (2019). https://doi.org/10.1016/j.fusengdes.2019.03.149

30. N. Taylor, S. Ciattaglia, D. Coombs, X.Z. Jin, J. Johnston, K. Liger, G. Mazzini, A. Widdowson, Safety and environment studies for a European DEMO design concept. Fusion Eng. Des. **146**(Part A), 111–114 (2019). https://doi.org/10.1016/j.fusengdes.2018.11.049

31. P. Chen, J. Tong, T. Liu, Solving the issue of reliability data for FOAK equipment in an innovative nuclear energy system. Progr. Nucl. Energy **163**, 104817 (2023). https://doi.org/10.1016/j.pnucene.2023.104817

32. B. Merk, D. Litskevich, A. Detkina, O. Noori-kalkhoran, L. Jain, E. Derrer-Merk, D. Afly-atunova, G. Cartland-Glover, Imagine—visions, missions, and steps for successfully delivering the nuclear system of the 21st century. Energies **16**, 3120 (2023). https://doi.org/10.3390/en1 6073120

33. H. Xu, B. Zhang, Diverse and flexible coping strategy for nuclear safety: opportunities and challenges. Energies **15**(17), 6275 (2022). https://doi.org/10.3390/en15176275

34. K. Kim, H.J. Kim, Numerical approach for cocurrent stratified steam water flow in a horizontal configuration. KSME J. **1**(2), 158–166 (1987). https://doi.org/10.1007/bf02971660

35. V.V. Kharchenko, O.Y. Chirkov, S.V. Kobel's'kyi, V.I. Kravchenko, A.O. Zvyagintseva, Application of refined calculation guidelines to the stress-strain state and fracture resistance analysis of the NPP primary-circuit system elements. Strength Mater. **53**(6), 824–833 (2021). https://doi.org/10.1007/s11223-022-00349-8

36. International Organization for Standardization, ISO 37000:2021 Governance of organizations—Guidance (2021). https://www.iso.org/standard/65036.html

37. International Atomic Energy Agency, Ageing management for research reactors. Specific safety guide (2023), in *IAEA Safety Standards Series*, No. SSG-10. (Rev. 1), STI/PUB/2050 (2010)

38. International Atomic Energy Agency, Ageing management for nuclear power plants. Specific safety guide, in *IAEA Safety Standards Series*, No. NS-G-2.12, STI/PUB/1373 (2009)

39. International Atomic Energy Agency, Ageing management and development of a programme for long term operation of nuclear power plants. Specific safety guide, in *IAEA Safety Standards Series*, No. SSG-48, STI/PUB/1814 (2018)

40. International Atomic Energy Agency, Periodic safety review for nuclear power plants. Specific safety guide, in *IAEA Safety Standards Series*, No. SSG-25, STI/PUB/1588 (2013)

41. International Atomic Energy Agency, Advanced control systems to improve nuclear power plant reliability and efficiency, in *IAEA-TECDOC-952* (1997), 186 p.

42. International Atomic Energy Agency, Fundamental safety principles IAEA, in *Safety Standards Series*, No. SF-1, STI/PUB/1273 (2006), 21 p.

43. International Electrotechnical Commission, IEC 60231:1967 «General principles of nuclear reactor instrumentation» by IEC TC/SC 45A (1967)

44. International Electrotechnical Commission, IEC 60568:2006 «Nuclear power plants—Instrumentation important to safety-In-core instrumentation for neutron fluence rate (flux) measurements in power reactors», TC 45/SC 45A (2006)

45. International Atomic Energy Agency, Safety of nuclear power plants: design/description, in *Series: IAEA Safety Standards Series* (International Atomic Energy Agency, Vienna, 2016)

46. International Atomic Energy Agency, Safety of research reactors: safety requirements, in *Safety Standards Series* (International Atomic Energy Agency, Vienna, 2005)

47. International Atomic Energy Agency, Safety of nuclear fuel cycle facilities, in *IAEA Safety Standards Series* (Vienna, 2014)

48. International Atomic Energy Agency, *Defence in Depth in Nuclear Safety: INSAG-10/a Report by the International Nuclear Safety Advisory Group* (INSAG Series, Vienna, 1996)

Correction to: Renewables in the Circular Economy and Business

Viktor Koval

Correction to:
V. Koval (ed.), *Renewables in the Circular Economy and Business*, SpringerBriefs in Applied Sciences and Technology, https://doi.org/10.1007/978-3-031-72174-8

In the original version of the book, the copyright years at the book were not correctly stated. This has now been rectified and the correct copyright year has been updated.

The updated version of this book can be found at
https://doi.org/10.1007/978-3-031-72174-8